Felix Stoehr

Simulations of Galaxy Formation and Large Scale Structure

Felix Stoehr

Simulations of Galaxy Formation and Large Scale Structure

PhD Thesis, Ludwig-Maximilians-Universität München (LMU)

Südwestdeutscher Verlag für Hochschulschriften

Impressum/Imprint (nur für Deutschland/ only for Germany)
Bibliografische Information der Deutschen Nationalbibliothek: Die Deutsche Nationalbibliothek verzeichnet diese Publikation in der Deutschen Nationalbibliografie; detaillierte bibliografische Daten sind im Internet über http://dnb.d-nb.de abrufbar.

Alle in diesem Buch genannten Marken und Produktnamen unterliegen warenzeichen-, marken- oder patentrechtlichem Schutz bzw. sind Warenzeichen oder eingetragene Warenzeichen der jeweiligen Inhaber. Die Wiedergabe von Marken, Produktnamen, Gebrauchsnamen, Handelsnamen, Warenbezeichnungen u.s.w. in diesem Werk berechtigt auch ohne besondere Kennzeichnung nicht zu der Annahme, dass solche Namen im Sinne der Warenzeichen- und Markenschutzgesetzgebung als frei zu betrachten wären und daher von jedermann benutzt werden dürften.

Verlag: Südwestdeutscher Verlag für Hochschulschriften Aktiengesellschaft & Co. KG
Dudweiler Landstr. 99, 66123 Saarbrücken, Deutschland
Telefon +49 681 37 20 271-1, Telefax +49 681 37 20 271-0
Email: info@svh-verlag.de
Zugl.: München, LMU, Diss., 2003

Herstellung in Deutschland:
Schaltungsdienst Lange o.H.G., Berlin
Books on Demand GmbH, Norderstedt
Reha GmbH, Saarbrücken
Amazon Distribution GmbH, Leipzig
ISBN: 978-3-8381-0884-1

Imprint (only for USA, GB)
Bibliographic information published by the Deutsche Nationalbibliothek: The Deutsche Nationalbibliothek lists this publication in the Deutsche Nationalbibliografie; detailed bibliographic data are available in the Internet at http://dnb.d-nb.de.

Any brand names and product names mentioned in this book are subject to trademark, brand or patent protection and are trademarks or registered trademarks of their respective holders. The use of brand names, product names, common names, trade names, product descriptions etc. even without a particular marking in this works is in no way to be construed to mean that such names may be regarded as unrestricted in respect of trademark and brand protection legislation and could thus be used by anyone.

Publisher: Südwestdeutscher Verlag für Hochschulschriften Aktiengesellschaft & Co. KG
Dudweiler Landstr. 99, 66123 Saarbrücken, Germany
Phone +49 681 37 20 271-1, Fax +49 681 37 20 271-0
Email: info@svh-verlag.de

Printed in the U.S.A.
Printed in the U.K. by (see last page)
ISBN: 978-3-8381-0884-1

Copyright © 2010 by the author and Südwestdeutscher Verlag für Hochschulschriften Aktiengesellschaft & Co. KG and licensors
All rights reserved. Saarbrücken 2010

PhD Thesis
4 July 2003

Physics Department
Ludwig-Maximilians-Universität München

Supervising Professor

Prof. Dr. Simon D. M. White
Max-Planck-Institut für Astrophysik, Germany

1. Gutachter: Prof. Dr. Simon D. M. White
2. Gutachter: Prof. Dr. Gerhard Börner

Tag der mündlichen Prüfung: 9. Oktober 2003

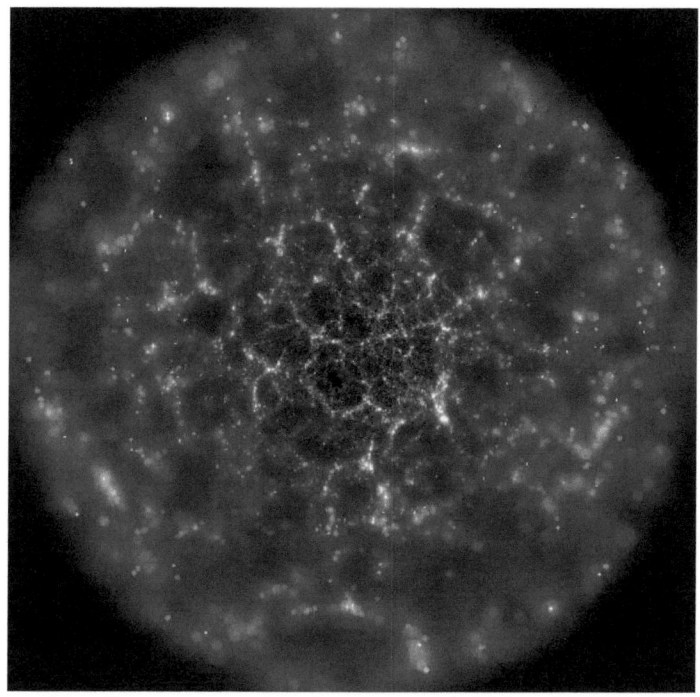

"Wenn die Vorstellung aller dieser Vollkommenheit nun die Einbildungskraft rührt, so nimmt den Verstand andererseits eine andere Art der Entzückung ein, wenn er betrachtet, wie so viel Pracht, so viel Grösse aus einer einzigen allgemeinen Regel mit einer ewigen richtigen Ordnung abfliesst."

Immanuel Kant, Allgemeine Naturgeschichte und Theorie des Himmels, 1755

Contents

Zusammenfassung		**9**
Summary		**11**
1 The Standard Model of Galaxy Formation		**15**
1.1	The Universe in a nutshell	16
1.2	Large scale structure	18
	1.2.1 The standard model of cosmology	18
	1.2.2 Dark matter	19
	1.2.3 The homogeneous Universe	21
	1.2.4 Density fluctuations	23
	1.2.5 Fluctuation growth	25
	1.2.6 Object formation	27
	1.2.7 Limits of the standard model	29
1.3	Galaxy formation	29
	1.3.1 Cooling	30
	1.3.2 Star formation	30
	1.3.3 Feedback	32
	1.3.4 Merging	32
	1.3.5 Reionisation	33
1.4	Numerical simulations	33
	1.4.1 Zel'dovich approximation	35
	1.4.2 Simulation codes	36
	1.4.3 Semi-analytic models	37
	1.4.4 DM-only vs. SPH	38
2 Galaxies and Environment		**39**
2.1	Introduction	40
2.2	N-body simulations	41
	2.2.1 Initial conditions	41

Contents

			2.2.1.1	Region selection	42
			2.2.1.2	Grid and glass distributions	45
			2.2.1.3	Simulations	46
	2.3	Halo environment			49
		2.3.1	Estimators		50
			2.3.1.1	Mass in shells	50
			2.3.1.2	FOF subtraction	50
			2.3.1.3	Comparison	52
	2.4	Dark Matter Properties			52
		2.4.1	Mass function		54
		2.4.2	Profiles		54
		2.4.3	Shapes		57
		2.4.4	Spin parameter		61
		2.4.5	Formation time		61
		2.4.6	Last major merger		66
	2.5	Galaxies			66
		2.5.1	Galaxy formation model		70
			2.5.1.1	Cooling	71
			2.5.1.2	Star formation	72
			2.5.1.3	Feedback	73
			2.5.1.4	Merging	73
			2.5.1.5	Model parameters	74
		2.5.2	Luminosity function		75
		2.5.3	Tully-Fisher-relation		75
		2.5.4	Star formation rate		76
		2.5.5	Galaxy distribution		79
		2.5.6	Galaxies and environmental effects		80
			2.5.6.1	Bulge-to-total luminosity ratio	80
			2.5.6.2	B-V colour index	85
	2.6	Conclusions			85
3	**Satellite Galaxies of the Milky Way**				**91**
	3.1	Introduction			92
	3.2	The simulated Milky Way			94
		3.2.1	Initial conditions		95
			3.2.1.1	Selection procedure	95
			3.2.1.2	Twolevel-ZIC	96
			3.2.1.3	Starting redshift	101

		3.2.1.4	Contamination 104

- 3.2.1.4 Contamination 104
- 3.2.1.5 Simulations 104
- 3.2.1.6 Scaling 105
- 3.2.2 Simulation results 107
  - 3.2.2.1 Density profile and circular velocity curve 107
  - 3.2.2.2 Bound vs. unbound subhalo particles 115
  - 3.2.2.3 Backtracking 116
  - 3.2.2.4 Circular velocity curves 123
  - 3.2.2.5 Object-by-object comparison 128
- 3.3 Observations 129
  - 3.3.1 Milky Way dwarf spheroidals 129
  - 3.3.2 Fornax and Draco 131
- 3.4 Velocity dispersion 133
  - 3.4.1 Phase space density 133
  - 3.4.2 The Jeans equation 134
- 3.5 Comparison 136
  - 3.5.1 Central velocity dispersions 136
  - 3.5.2 Velocity dispersion profiles 138
- 3.6 Conclusions 141

4 Dark Matter Annihilation 143
- 4.1 Introduction 144
- 4.2 Dark matter 145
  - 4.2.1 Candidate particles 145
  - 4.2.2 Supersymmetry 146
  - 4.2.3 Dark matter search 146
  - 4.2.4 Annihilation 148
- 4.3 N-body simulations 149
- 4.4 Smooth halo luminosity 149
- 4.5 Halo substructure 153
  - 4.5.1 Subhalomassfunction 153
- 4.6 Substructure luminosity 154
  - 4.6.1 Density estimation 155
  - 4.6.2 Enhancement due to substructure 156
    - 4.6.2.1 Flattening and discreteness 156
    - 4.6.2.2 Selfsimilarity 158
    - 4.6.2.3 Luminosity contributions 160
  - 4.6.3 Results 161

Contents

4.7	Subhalodistribution	163
4.8	Detectability	163
	4.8.1 Cross-section computation	166
	4.8.2 Background	169
	4.8.3 Results	170
4.9	Conclusions	173

Conclusions   **175**

List of Publications   **179**

List of Figures   **181**

List of Tables   **185**

Units and Constants   **187**

Bibliography   **189**

Zusammenfassung

Das Studium der Entstehung und Entwicklung von Galaxien ist eines der interessantesten Gebiete der Kosmologie. In dieser Arbeit verwenden wir die sogenannte Resimulationstechnik und berechnen ultrahochauflösende Computersimulationen der Strukturentstehung im Universum.

Zunächst simulieren wir eine sorgfältig ausgewählte große Region mittlerer Dichte mit vier verschiedenen Auflösungen und untersuchen die Abhängigkeit der Eigenschaften der Halos aus dunkler Materie von der Materiedichte in ihrer lokalen Umgebung. Wir finden eine schwache Abhängigkeit der Spin- und Konzentrationsparameter. Die Form der Halos, ihre Entstehungszeitpunkte sowie der Zeitpunkt des letzten Verschmelzens mit einer etwa gleich großen Galaxie sind dagegen umgebungsunabhängig. In einem weiteren Schritt wenden wir ein semi-analytisches Modell der Galaxienentstehung auf die Simulationen an. Unsere Modellgalaxien zeigen starke Umgebungsabhängigkeiten.

Im zweiten Teil dieser Arbeit vergleichen wir die beobachtete Struktur und Kinematik der bekannten Satellitengalaxien der Milchstraße mit den Eigenschaften simulierter Subhalos. Es war behauptet worden, diese beobachteten Eigenschaften würden nicht mit den simulierten übereinstimmen, was das kosmologische ΛCDM Modell in Frage gestellt hat. Wir berechnen die Entstehung des Halos einer Galaxie von der Größe der Milchstraße mit bisher unerreichter Auflösung und untersuchen dieses sogenannte "Substruktur-Problem" erneut. Wir finden, daß die von unserer Simulation vorhergesagten Geschwindigkeitsdispersionen beobachteter Sternverteilungen exakt mit den beobachteten Dispersionen übereinstimmen. Dies ist ein Triumph für das ΛCDM Modell. Andere Formen dunkler Materie, die vorgeschlagen wurden, um das Substruktur-Problem zu lösen, würden die gefundene Übereinstimmung zerstören.

Im dritten Teil dieser Arbeit untersuchen wir, ob die Gammastrahlung, die durch Selbst-Annihilation entstehen kann, falls die dunkle Materie im Universum aus schwach wechselwirkenden massiven Teilchen besteht, mit der nächsten Generation von Gammastrahlungsteleskopen beobachtet werden kann. Wir verwenden unsere hochauflösende Simulation des Galaxienhalos und finden, daß frühere Untersuchungen die Bedeutung der Gammastrahlung von Subhalos überschätzt haben. Wir schlagen eine neue Beobachtungsstrategie vor und zeigen, daß bei derzeit diskutierten Modellen für die Teilchen der dunklen Materie das galaktische Zentrum mit GLAST beobachtet werden kann.

Zusammenfassung

Summary

Galaxy formation is one of the most fascinating topics of modern cosmology. Since time immemorial, people have desired to understand the origin, motion and evolution of planets, stars and, more recently, galaxies and the Universe as a whole. Great advances in astronomy always have had impact on philosophy and redefined the self-understanding of mankind within the Universe.

The first milestone on the long road of discoveries was undoubtedly the formulation of the laws of gravity and mechanics in 1687 by Newton. Einstein's extension of these laws in the years 1905 and 1913 led to a revolutionised understanding of space and time. In 1929 Hubble established the expanding Universe which subsequently led to the postulation of the hot Big Bang by Lemaître (1934). Zwicky (1933) found that most of matter in the Universe is dark. The nature of this matter, interacting only through gravity and perhaps through the weak interaction, is still a mystery. Finally, Penzias and Wilson (1965) discovered the cosmic microwave background radiation, not only confirming the theory of the Big Bang, but also - as was observed later - revealing the origin of structure in the Universe.

Today, cosmology and especially galaxy formation are fast paced exciting scientific fields. Surveys like the Sloan Digital Sky Survey will soon provide a catalogue of about 500 million galaxies with an unprecedent wealth of data. Deep observations with 8 or 10 m telescopes or with the Hubble Space telescope allow to observe objects in their very early evolutionary stages.

In addition to this, the dramatic increase in computer power now allows us to carry out numerical experiments on galaxies and even on the large scale structure of the Universe. The latter is possible because of the extraordinary fact that as a result of microwave background observations the properties of the Universe some 300,000 years after the Big Bang are well known.

As ordinary matter makes up only about ten percent of the total matter in the Universe, it can be neglected in simulations in a first approximation. An initial density field can then be evolved under the sole influence of gravity. The result of such simulations may be combined with semi-analytic models for the baryonic physics associated with galaxy formation.

Summary

Gravity is a long range force, and it turns out that length scales of 100 Mpc or more have to be included in large scale structure simulations in order to obtain results that are representative for the Universe as a whole. The sizes of galaxies, however, are three to four orders of magnitude smaller than this so that numerical resolution has always been a concern in simulations which try to include galaxy formation.

A clever and powerful trick alleviates this problem. After a low-resolution simulation has been performed, a small region of interest is selected and the simulation is run again, this time concentrating most of the computational effort on the small region, allowing the resolution to be increase dramatically without losing tidal influences from the large cosmological volume.

This technique - called resimulation - is the driving force behind all the simulations that were performed for this thesis. After having run about 1500 supercomputer jobs it is clear that this technique is extremely powerful and allows the faithful simulation of objects that are far into the regime of non-linear evolution while taking into account the full cosmological context.

In the first chapter of this work we briefly introduce aspects of the observable Universe and discuss the relevant theoretical background for this thesis.

In the second chapter we use high-resolution simulations of structure formation to investigate the influence of the local environment of dark matter haloes on their properties. We run a series of four re-simulations of a typical, carefully selected representative region of the Universe so that we can explicitly check for convergence of the numerical results. In our highest resolution simulation we are able to resolve dark matter haloes as small as the one of the large Magellanic cloud.

We propose a new method to estimate the density in the environment of a collapsed object and find weak correlations of the spin parameter and the concentration parameter with the local halo density. We find no such correlation for the halo shapes, the formation time and the last major merging event. In a second step we produce catalogues of model galaxies using a semi-analytic model of galaxy formation. We find correlations between the bulge-to-disk luminosity and the B-V colour index with the local environment.

In chapter three we compare observations of the internal structure and kinematics of the eleven known satellites of the Milky Way with simulations of the formation of its dark halo in a ΛCDM universe. Earlier work by Moore et al. (1999) and Klypin et al. (1999) claimed the cosmological concordance model of the Universe, the ΛCDM model, to disagree with observations. The so-called "substructure-problem" is one of the two major challenges for this model and has attracted much attention. In order to remove the discrepancy, changes of the cosmological model have been proposed. We reinvestigate the substructure-problem using our ultra-high resolution simulations. For a galaxy-sized dark matter halo, our mass resolution is the highest resolution ever achieved. In contrast to the work of Moore et al. (1999) and Klypin et al. (1999), we find excellent agreement. The observed kinematics are exactly those predicted for stellar populations with the observed spatial structure orbiting within the most massive "satellite" substructures in our simulations. Less massive substructures have weaker potential wells than those hosting the observed satellites. If there is a halo substructure "problem", it consists in understanding why halo substructures have been so inefficient in making stars. We find that suggested modifications of dark matter properties (e.g. self-interacting or warm dark matter) may well spoil the good agreement found for standard Cold Dark Matter.

If the dark matter in the Universe is made of weakly self-interacting particles, they may self-annihilate and emit γ-rays. The detection of the γ-ray signal would finally, after seventy years since its discovery, shed light on the nature of the dark matter. In chapter four we use our ultra-high resolution numerical simulations to estimate directly the annihilation flux from the central region of the Milky Way and from dark matter substructures in its halo. Such estimates remain uncertain because of their strong dependence on the structure of the densest regions of the halo. Our numerical experiments suggest, however, that less direct calculations have typically overestimated the emission from the centre of the Milky Way and from its halo's substructure. We find an overall enhancement of at most a factor of a few with respect to a smooth halo of standard NFW structure. For an observation outside the region around the galactic centre where the diffuse galactic γ-ray background is dominant, GLAST can probe a large region of possible MSSM models. This result is independent of the exact structure of the innermost region of the Galaxy. Our analysis shows that the flux from the inner galaxy exceeds the expected contribution from the brightest substructure by a large factor. Nevertheless, for certain MSSM models substructure halos might be detectable with GLAST.

Summary

1 The Standard Model of Galaxy Formation

1. The Standard Model of Galaxy Formation

In this chapter we first give a very short overview of the observable Universe before we briefly describe the standard model of cosmology and the theory of structure formation. In section 1.3 we discuss the main processes relevant for galaxy formation in this framework and finally outline the background that is necessary for understanding the simulations studied in the following chapters.

1.1 The Universe in a nutshell

Our planet Earth is orbiting together with eight other planets around the Sun. The Earth has a mass of about 6.0×10^{24} kg, i.e. 3×10^{-6} $M_\odot$ (see appendix D for a summary of the units used in this work) and has formed about 4.7 Gyr ago. The Sun has a mass of 2.0×10^{30} kg, and is located at 1.5×10^{11} m (i.e. 4.8×10^{-6} pc) from us. Sun and planets form the solar system which has a diameter of about 1.2×10^{13} m (i.e 3.4×10^{-4} pc). The nearest star to the Sun is Proxima Centauri at 1.3 pc. Probably about two thirds of the stars are bound in binary systems.

The stars (together with their planets) are themselves grouped in galaxies. The galaxies are given classes according to their types (irregular, spiral, lenticular and elliptical galaxies) and sizes (from dwarf spheroidal to giant elliptical galaxies). The smallest galaxies contain about 10^5 stars whereas the largest ones have up to 10^{13} stars. Most of the galaxies, however, are made up of about 10^{10} to 10^{12} stars and have diameters of about 10 to 30 kpc.

Galaxies can be split in two major classes: disk and elliptical galaxies. Elliptical galaxies are smooth, featureless, ellipsoidally shaped and contain mostly old stars. These stars are moving on random orbits. Disk galaxies contain young stars, much gas and dust. They often show spiral arm and bar features. The disks are rotating and are rotationally supported.

The different stellar populations of galaxies give valuable clues on their formation histories. The current understanding is that the different galaxy types are stages of their evolution from the collapse of clouds of primordial gas, through matter accretion and galaxy-galaxy merging. Whereas stellar encounters are extremely rare, mainly due to their small sizes compared to the mean separation, galaxy collisions are believed to have happened quite frequently: Elliptical galaxies are believed to have formed through mergers of disk galaxies of similar mass. The build up of the structures in the Universe is thought to have happened in a hierarchical manner. Larger structures are formed through the merging of smaller ones.

Our own galaxy, the Milky Way (MW), is a spiral galaxy. It contains about 2×10^{11} stars which are orbiting in a disk of radius 18 kpc and thickness of 300 pc as well as in a spheroidal bulge of radius 1.8 kpc. A black hole of mass 2.5×10^6 $M_\odot$ is at the very

centre. The Sun is orbiting in the outskirts of the disk at about 8 kpc from the centre with a velocity of about 220 km/s. It completes one full revolution every 0.22 Gyr.

The galaxies themselves are, just like the stars, distributed inhomogeneously. They come in entities from a few galaxies, called groups, to thousands of galaxies, called galaxy clusters. Our own galaxy is the second largest galaxy of the Local Group (LG) which contains about 30 galaxies. The largest LG object is M31 (Andromeda) which is a spiral galaxy just like the MW. The two galaxies are approaching at about 70 km/s due to their gravitational attraction. The LG itself is moving with about 620 km/s with respect to the local restframe (see next section). The closest galaxy cluster to the MW is Virgo at 60 Mpc containing about 2000 galaxies. In contrast to Virgo, the galaxies in Coma are concentrated towards the centre. Coma, situated at roughly 90 Mpc from the MW is about four times more massive.

Galaxy clusters are the largest quasi-equilibrium systems in the Universe. The typical separation of galaxy clusters is of the order of 50 Mpc. The distribution of the galaxy clusters, groups and galaxies forms a web-like structure with clusters and filaments surrounding large underdense regions, called voids (see Figure 2.2).

The different types of galaxies are distributed differently in the Universe. Whereas spiral galaxies make up nearly eighty per cent of the galaxy population in the low density regions of the Universe, they contribute with only ten per cent in galaxy clusters (Hubble & Humason (1931); G.C. McVittie (1962); Dressler (1980)).

The latest measurements suggest that the Universe, since the Big Bang, has an age of 13.4 Gyr. This can be converted into the radius of the observable part of 13.7 Gpc. Due to the finite value of the speed of light, observations of distant objects show them in an early state of their evolution. Although observations are only a snapshot of the evolution of the Universe, observing distant objects allows to get information about the structure and galaxy formation processes.

The galaxy distribution is not static, but on average, galaxies are receding faster from the MW the more distant they are (Hubble (1929)). The frequencies of photons emitted at earlier times are shifted to lower values, i.e. to the red.

$$z = \frac{\lambda_{obs}}{\lambda_{lab}} - 1 = \frac{\nu_{lab}}{\nu_{obs}} - 1 \qquad (1.1)$$

The subscripts denote the quantities observed and measured in the laboratory, respectively. This redshift can be explained with the expansion of the Universe (section 1.2.3). The object with the highest redshift discovered so far is the galaxy HCM 6A in the Subaru Deep Field at $z = 6.58$ (Hu et al. 2002).

Using galaxies as measurement units, one finds that the distribution on the sky (2D and nowadays also 3D) is approximately isotropic and homogeneous on scales larger than

1. The Standard Model of Galaxy Formation

about 100 Mpc. A subregion of the Universe must therefore be as large or larger than this in order to be representative of the Universe as a whole.

Studies of the velocities of galaxies in galaxy clusters or stars in galaxies revealed that the mass of the stars is by far too small to be the source of the gravitational potential the galaxies or stars orbit in (Zwicky (1933); Peacock (1993) and references therein). Only a small fraction, about 1.5 % of the matter in the universe can actually be seen. The average density of this visible matter is about 10^{-31}g/cm^3.

1.2 Large scale structure

The ultimate task of cosmology is to find a theory describing the observations just presented. Although the final theory is not in reach today, enormous progress has been made in the recent years and it is possible to claim that a "standard model of structure and galaxy formation" has emerged. In this section we give the background of structure formation relevant for our work. The processes relevant for galaxy formation are treated in the next section.

1.2.1 The standard model of cosmology

The standard model is based on four pillars. The first one is the theory of the hot Big Bang. Georges Lemaître was the first to realize that following the expansion of the Universe back in time, the initial state must have been of very high density and temperature. After the Planck time (i.e., $t > 10^{-43}$ s) the Universe started to expand and the temperature dropped. For the time before the Planck time a theory of quantum gravity is required. Finding such a theory is to date one of the major activities in modern physics. As the definite theory has yet to be found, the description of the initial state of the Universe is not possible. The current understanding is that the Universe underwent several phase transitions that led to the appearance of the physical forces. Between about 10^{-11} seconds and 100 seconds after the Big Bang the temperature was low enough so that particles (electrons, quarks, protons, neutrons, Weakly Interacting Massive Particles (WIMPS), baryons etc.) froze out: When the reaction time scales become longer than the expansion time scale the creation and destruction reactions of a particle species are not in equilibrium any more. For stable particles the abundance becomes constant. Around $t = 10000$ yr, most of the energy density was not in radiation any more; the Universe became matter dominated and the primordial composition of particles was fixed (see also next section).

At $t = 300,000$ yr, free electrons could be captured by protons and α-particles transforming the ionised plasma into neutral atoms. The mean free path of the thermalized

photons, previously constantly scattered in the plasma, increased dramatically. The radiation, the famous microwave background, has a temperature of 2.728 ± 0.004 K today (Spergel et al. 2003). It has a perfect black body radiation spectrum. Although discovered only in 1965 (Penzias & Wilson 1965), the microwave background had been predicted already earlier (Gamov 1946; Alpher, Bethe & Gamov 1948).

After subtraction of the dipole anisotropy due to the motion of the Earth with respect to the local rest-frame (371 ± 1 km/s), the background radiation is isotropic to better than 10^{-4} on scales larger than about 12 arcminutes! Although previous observations already showed isotropy on large scales, the confirmation from the microwave background is highly impressive. With the reasonable assumption that the Universe is isotropic also seen from a different position, the second pillar of the standard model, the assumption of the homogeneous and isotropic Universe (on large scales) can be confirmed. We will briefly discuss the consequences of the tiny deviations from the isotropy below.

The third pillar of the standard model of cosmology is the assumption that the theory of general relativity describes gravitation and space time geometry correctly. For the homogeneous, isotropic Universe, Einstein's equations can be written in a rather simple form. So far, the theory has passed all tests (van Straten et al. (2001) and references therein). Although improbable, the possibility remains that the law of gravity is not newtonian for very small accelerations (Milgrom 1983; Sanders & McGaugh 2002).

The theory of the nucleosynthesis during the evolution of the early Universe is very mature today. The predictions of the abundances of the light elements (H, ^{2}D, ^{3}He, ^{4}He and ^{7}Li) are consistent with the observed abundances over 10 orders of magnitude. This theory can be considered the fourth pillar of the cosmological standard model. The best fit value for the abundance of baryons in the Universe is in very good agreement with the latest determinations from the cosmic microwave background (CMB) $\Omega_b = \rho_m/\rho_c = 0.047 \pm 0.006$. Where the critical density ρ_c

$$\rho_c = \frac{3H_0^2}{8\pi G} \qquad (1.2)$$

is the density required for a closed Universe (section 1.2.3). This is a very strong argument, as the two observations are independent. The theory of the Big Bang nucleosynthesis (BBN) is so predictive because for each of the light elements the prediction can be compared to the observations independently. The free parameters of the model, however have to be the same for all of them.

1.2.2 Dark matter

As mentioned in section 1.1, most of the matter in the Universe does not emit radiation and manifests itself only via gravity and perhaps weak interaction. This was first estab-

1. The Standard Model of Galaxy Formation

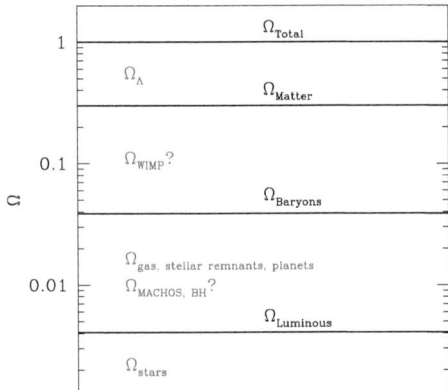

Figure 1.1: Contributions of the different forms of matter to the total matter density. The values are given for the current concordance ΛCDM model. The luminous matter makes up only 0.4% of the total matter density. On all scales there are unknown matter or energy components. The left Ω-labels indicate the probable cosmological objects that provide the missing matter densities.

lished by Fritz Zwicky 1933, on the scale of galaxy clusters. Today there is evidence for Dark Matter on all mass scales from galaxy clusters to dwarf galaxies (see also chapters 3 and 4).

The total amount of visible matter is about $\Omega_{vis} \approx 0.004$. Together with the measurement from BBN this means that about 90 % of the baryons are actually "dark". Observations of the Lyman-α-forest suggest that these baryons form molecular gas of low column-density.

The determination of the *total* matter density in the Universe has been a longstanding task in cosmology. The best value today is $\Omega_m = 0.29 \pm 0.07$ (Spergel et al. 2003). This leads to an astonishing result: about 90% of the matter in the Universe is made of non-baryonic, as yet unknown, particles! Many candidate particles have been proposed (neutrinos, axions, WIMPS, ...). They are regrouped in the categories "hot", "warm" and "cold", for light (typically eV), medium (typically keV) and heavy (typically MeV

to TeV) particles. Hot DM particles decouple when they are still relativistic. They are roughly as abundant as photons and due to their high velocities can traverse large regions in the early Universe. The CMB density fluctuations on small scales (see below) will then be washed out. Cold DM particles decouple when they are non-relativistic. CMB density fluctuations would remain intact down to small scales due to the nearly vanishing thermal velocities. Warm DM particles are an intermediate case. Although the particles decouple when they are still relativistic, their abundance is suppressed.

As for now, only hot dark matter is ruled out (Hut & White 1984). A warm DM particle must probably have a rest-mass much higher than 10 keV in order to not violate the latest CMB measurements (Yoshida et al. 2003). The most probable candidate for the DM particle is the WIMP (see also chapter 4).

For the remainder of this work we will use the term "DM" only for the non-baryonic DM component. Processes involving baryons will be discussed in section 1.3 and chapter 2.

1.2.3 The homogeneous Universe

For the very special case of a homogeneous and isotropic Universe the equations of general relativity become rather simple. This is not astonishing because the properties of space time depend on the matter distribution. As no position or direction are privileged, any observer at rest with respect to the matter in the local frame will measure the same proper time. This proper time is an universal time coordinate. It is convenient to introduce *comoving coordinates* **r**

$$\mathbf{x}(t) = a(t)\mathbf{r}(t) \tag{1.3}$$

where the general expansion of the Universe $a(t)$ has been factored out. The usual convention is that the value of the present day expansion factor $a(t = t_0)$ is set to 1.

The most general metric for a homogeneous and isotropic Universe is the Robertson-Walker (RW) metric (see Peacock (1993)) in spherical coordinates (r, θ, ϕ):

$$c^2 d\tau^2 = c^2 dt^2 - a^2(t) \left(\frac{dr^2}{1 - kr^2} + r^2 (d\theta^2 + \sin^2(\theta)\, d\phi^2) \right). \tag{1.4}$$

Depending on the value of k the Universe is "flat" ($k < 0$), "open" ($k = 0$) or "closed" ($k > 0$). These notations correspond to Universes in which the space time is flat, negatively and positively curved, respectively. Note that for the case of a flat Universe as observations suggest (Spergel et al. 2003), the metric simplifies even further.

With the RW metric (equation 1.4), Einstein's equations become

$$\frac{\ddot{a}}{a} = -\frac{4\pi G}{3}\left(\rho + \frac{3p}{c^2}\right) + \frac{\Lambda c^2}{3} \tag{1.5}$$

1. The Standard Model of Galaxy Formation

and
$$\left(\frac{\dot{a}}{a}\right)^2 = \frac{8\pi G}{3} - \frac{kc^2}{a^2} + \frac{\Lambda c^2}{3} \quad (1.6)$$

Here, ρ is the matter density, p the pressure and Λ the cosmological constant. This constant corresponds to an energy density of the vacuum

$$\rho_v c^2 = \frac{\Lambda c^4}{8\pi G} = \Omega_\Lambda \rho_c c^2 \quad (1.7)$$

Note that we omit the subscripts "$_0$" for present-day values on the density parameters Ω as well as on the cosmological constant Λ.

The expansion rate of the Universe and the expansion factor are related by Hubble's law, i.e.
$$H(a) = \frac{\dot{a}}{a}. \quad (1.8)$$

At present time, the proportionality constant H_0, the *Hubble constant*, has a value of 72 ± 8 km/s/Mpc (Freedman et al. 2001). It is usual to introduce a dimensionless constant h defined as $H_0 = 100\, h$ km/s/Mpc.

Photon wavelengths stretch with the expansion if the Universe. The spectral lines get redshifted according to
$$z + 1 = \frac{1}{a}. \quad (1.9)$$

The second Friedmann equation can now be transformed into
$$\frac{H^2(z)}{H_0^2} = \Omega_m(1+z)^3 + \Omega_c(1+z)^2 + \Omega_\Lambda. \quad (1.10)$$

We here have introduced the density parameter of the curvature of space time
$$\Omega_c = -\frac{k\, c^2}{H_0^2}. \quad (1.11)$$

Given that we know the values of the present-day density parameters we can integrate equation 1.10 using equations 1.8 and 1.9

$$t = \int_0^z \frac{dx}{(1+x)H_0\sqrt{\Omega_m(1+x)^3 + \Omega_c(1+x)^2 + \Omega_\Lambda}} \quad (1.12)$$

This equation relates the cosmological redshift to the look back time and is a very useful equation in cosmology. The comoving distance d to an object is

$$d = \int_0^z \frac{c\, dz'}{H(z')} \quad (1.13)$$

1.2 Large scale structure

When setting $z = \infty$ the lookback time t corresponds to the age of the Universe. For some combinations of the cosmological parameters there are analytical solutions to equation 1.12. For a flat space time and a non-vanishing cosmological constant this analytic solution is

$$t = \frac{2}{3H_0(1-\Omega_0)^{1/2}} \text{ arssinh}\left[\frac{\sqrt{1/\Omega_m - 1}}{(1+z)^{3/2}}\right] \tag{1.14}$$

To complete this section, we list the current best estimates of the cosmological parameters as determined by the WMAP collaboration.

Ω_m	0.29 ± 0.07
Ω_Λ	0.69 ± 0.05
Ω_c	0 (assumption)
Ω_b	0.0047 ± 0.006
H_0 [km/s/Mpc]	72 ± 5
age [Gyr]	13.4 ± 0.3
z_{reion}	17 ± 5
σ_8	0.9 ± 0.1
A	0.9 ± 0.1
n	0.99 ± 0.04

Table 1.1: Most recent values of the cosmological parameters as determined by the CMB experiment WMAP (Spergel et al. 2003). The value of H_0 is from Freedman et al. (2001).

1.2.4 Density fluctuations

As mentioned in the previous section, the microwave background radiation is not absolutely smooth but after the subtraction of the dipole anisotropy small temperature fluctuations remain. These fluctuations were studied by the satellite mission COBE and subsequently by BOOMERANG, ARCHEOPS, and WMAP. COBE and WMAP covered the whole sky. Decomposing the temperature fluctuations in spherical harmonics, it is possible to recover the underlying fluctuation spectrum. The measurement of these fluctuations is currently the best method to obtain the cosmological parameters.

At the origin of the observed temperature fluctuations were density fluctuations in the very early Universe. The currently most widely accepted model is that these fluctuations

1. The Standard Model of Galaxy Formation

were then expanded on very large scales by a period ($\Delta t \approx 10^{-33}$ s) of exponential expansion of the Universe. This period is called *inflation* and was first suggested by Guth (1981). The inflation scenario successfully eliminates severe problems of the standard model of cosmology (see section 1.2.7). How inflation started or stopped or which of the different inflationary theories proposed is the best one is still unknown.

The results from (Spergel et al. 2003) are consistent with a random gaussian scale free Harrison-Zeldovich initial fluctuation spectrum, i.e. $P_{initial}(k) \propto k^n$ with $n = 1$.

$$P(k,t) = P(\mathbf{k},t) = \langle |\delta(\mathbf{k},t)|^2 \rangle \qquad (1.15)$$

$$\delta(\mathbf{k},t) = \frac{1}{(2\pi)^3} \int d\mathbf{x}\ \delta(\mathbf{x},t)\ e^{-i\mathbf{kx}} \qquad (1.16)$$

During the evolution of the early Universe until the end of the epoch of recombination, the density fluctuations are modified. For a calculation of the final fluctuation spectrum, the plasma and the dark matter as well as the neutrinos and the photons have to be taken into account. The density fluctuations grow by self-gravitation. This growth is reduced by radiation pressure and small-scale fluctuations are dissipated by free streaming.

The cumulative effect is put into the transfer function $T(k)$ that can be computed, for example with the code CMBFAST (Seljak & Zaldarriaga 1996) which is publicly available. Analytical fits, for example from Bond & Efstathiou (1987) are also available:

$$T_{BE}(k) = \frac{1}{(1 + (6.4\ q + (3\ q)^{1.5} + (1.7\ q)^2)^{1.13})^{1/1.13}} \qquad (1.17)$$

where $q = k/(\Gamma\ h)$. The powerspectrum is then

$$P(k) = A\ T^2(k)\ k^n \qquad (1.18)$$

The normalisation constant A has to be determined either from the CMB directly (see Table 1.1) or with a measurement of the present-day fluctuation amplitude on a given scale. Usually the density contrast in spheres of 8 Mpc σ_8 is used (White, Efstathiou & Frenk 1993; Eke, Cole & Frenk 1996; Viana & Liddle 1996).

The fluctuations observed in the microwave background are a result of the density fluctuations: The photons that have to climb out of the potential wells lost energy and thus regions on the sky corresponding to overdensities have lower CMB temperatures. In addition to this effect, which is called Sachs-Wolfe effect, anisotropies arise due to the compression of radiation in high density regions and due to the Doppler shift resulting from the motion of the plasma.

1.2 Large scale structure

Finally, after recombination, the photon energy can be changed by the accumulated gravitational potential, by scattering by free electrons either in clusters (Sunyaev-Zeldovich effect) or in the (reionised) inter-galactic medium (IGM). These anisotropies are called *secondary anisotropies*.

As different cosmological parameter sets can yield very similar CMB power spectra, it is crucial to reduce the measurement error bars. A promising field for CMB measurements is the region of high multipole numbers and polarisation fluctuations which will allow to break some degeneracies.

1.2.5 Fluctuation growth

After the epoch of recombination, the density fluctuations grow purely by self-gravitation. We now give the formulae for the evolution of a weakly perturbed density field in an expanding Universe to first order approximation starting from the continuity equation (due to the conservation of mass), the Euler equation (the equation of motion in a gravitational potential) and the Poisson equation (the relation between a density field and the generated gravitational potential).

If the length scale of the perturbations is smaller than the effective cosmological horizon $d_H = c/H_0$, a Newtonian treatment of the evolution is valid. If in addition the mean free path of a particle is small, matter can be treated as an ideal fluid and the Newtonian equations governing the motion of gravitating particles in an expanding universe can be written in terms of comoving coordinates $(\mathbf{r}, \mathbf{u})$ (equation 1.3) with physical velocity

$$\mathbf{v} = \dot{\mathbf{x}} = \dot{a}\mathbf{r} + a\dot{\mathbf{r}} = H\mathbf{x} + a\mathbf{u} \tag{1.19}$$

where we introduced $\mathbf{u} := \dot{\mathbf{r}}$ the comoving peculiar velocity corresponding to the departures from the general expansion. The latter, corresponding to the term $H\mathbf{x}$, is called the *Hubble flow*. Dots indicate derivation with respect to the proper time of an observer following the particle's trajectory.

In comoving coordinates (without pressure term and with vanishing cosmological constant) the continuity equation, the Euler equation and the Poisson equation are (Peebles 1980; Peacock 1993)

$$\frac{d}{dt}\rho + \nabla(\rho \mathbf{u}) = 0 \tag{1.20}$$

$$\frac{d\mathbf{u}}{dt} = -\nabla\Phi \quad \text{and} \tag{1.21}$$

$$\Delta\Phi = 4\pi G(\rho - \rho_{mean}) \tag{1.22}$$

1. The Standard Model of Galaxy Formation

respectively where d/dt is the total time derivative

$$\frac{d}{dt} = \frac{\partial}{\partial t} + \mathbf{u} \circ \boldsymbol{\nabla} = \frac{\partial}{\partial t} + v_x \frac{\partial}{\partial x} + v_y \frac{\partial}{\partial y} + v_z \frac{\partial}{\partial z}. \quad (1.23)$$

We write the density field as a homogeneous mean density part with superposed density fluctuations

$$\rho(\mathbf{r}) = \rho_{mean}(1 + \delta(\mathbf{r})). \quad (1.24)$$

Equation 1.24 can now be introduced into equations 1.20, 1.21 and 1.22. The arising second derivative can be replaced using the Friedmann equation 1.6. The resulting equations are describing the evolution of density fluctuations in an expanding Universe under self-gravity.

$$\dot{\delta} + (1+\delta)\nabla \dot{\mathbf{r}} = 0 \quad (1.25)$$

$$\ddot{\mathbf{r}} + 2\frac{\dot{a}}{a}\dot{\mathbf{r}} = -\frac{1}{a^2}\nabla \Phi_p \quad (1.26)$$

$$\Delta \Phi_p = 4\pi G \rho_{mean} a^2 \delta \quad (1.27)$$

The peculiar gravitational potential used here

$$\Phi_p = -G \int d\mathbf{r}' \frac{\rho(\mathbf{r}') - \rho_{mean}}{|\mathbf{r} - \mathbf{r}'|} \quad (1.28)$$

corresponds to the part generated by the density *fluctuations*.

When the fluctuations δ are small, the equations simplify. Combining equations 1.10, 1.25, 1.26 and 1.27 and neglecting terms of higher order than linear, the equation of gravitational amplification of density perturbations can be obtained as

$$\ddot{\delta} + 2\frac{\dot{a}}{a}\dot{\delta} = 4\pi G \rho_{mean} \delta \quad (1.29)$$

There are two general solutions for equation 1.29 $\delta(z) \propto D(z)$: a decaying mode $D_d(z) = H(z)$ and a growing mode

$$D_g(z) = D_d(z) \int_z^\infty dx \frac{1+x}{D_d^3(x)} \quad (1.30)$$

which is the one relevant for structure formation.

For a given set of cosmological parameters the Friedmann equation 1.6 can be solved to obtain the solution for the expansion factor $a(t)$. The growth of initial density perturbations $\delta(\mathbf{x}, z) = \delta_0(\mathbf{x}) D_g(z)/D_g(z_0)$ can then be computed with equation 1.29. This works fine as long as the density fluctuations are small $\delta \ll 1$. When expressing the initial density field in its Fourier components $\delta(\mathbf{k})$ it becomes clear that each mode evolves independently from one another. The initial spectrum of the fluctuation will - in this regime - not change except that the amplitude of the fluctuations will grow with the growth factor from equation 1.30:

$$P(k,z) = P_0(k) \; T^2(k) \; \frac{D_g(z)}{D_g(z_0)} \qquad (1.31)$$

In the non-linear regime, power from the small scales is transfered to the larger scales. An analytical prediction of the evolution of the power spectrum in the non-linear regime has been given by Peacock & Dodds (1996). For parts of the evolution of a single density peak in the non-linear regime a few analytical solutions are available (see next section). For the complete treatment of the initial density field numerical simulations are required (see section 1.4).

1.2.6 Object formation

The density peaks expand and amplify under self-gravity until a critical overdensity threshold is reached. In the model of a spherical perturbation with homogeneous density, this threshold (for a $\Omega_m = 1$ model) is

$$\delta = \frac{9\pi^2}{16} \approx 5.55 \qquad (1.32)$$

at which time the corresponding linearly extrapolated density contrast is about 1.06 (Peebles 1980; White 1993; Peacock 1993). The spherical perturbation then breaks away from the general expansion and collapses. The collapsing matter elements exchange energy and the system virialises.

This process, the *violent relaxation* process, is very difficult to compute analytically. It is therefore not obvious to determine beforehand the density profile of the remaining object. Using the virial theorem, one can estimate that the overdensity of the final object will be of the order of one to three hundred depending on the assumptions.

N-body simulations show that the virialised part of the collapsed haloes, i.e. the approximately spherical region in which the average radial motions are close to zero indeed has a mean overdensity of about 200 (Cole & Lacey 1996). Throughout this work we will adopt however a value of 200 times the critical density to define the boundaries of collapsed objects as is common practice.

Much effort has been undertaken to calculate a priori the density profile of DM haloes (see Peacock (1993), section 3.2.2.1) giving solutions for simplified assumptions. The currently best estimate however comes from N-body simulations. Navarro, Frenk & White (1997) (hereafter NFW) find the profile to be universal with the shape (see also

1. The Standard Model of Galaxy Formation

section 3.2.2.1)

$$\rho(r) \sim \frac{1}{r\,(1+r)^2} \qquad (1.33)$$

This finding was confirmed (for example Tormen, Bouchet & White (1997), Power et al. (2003), section 3.2.2.1). There is however still some debate about the exact slope in the very inner part of the DM haloes (Moore et al. 1998; Ghigna et al. 2000) (see also section 3.1).

The number of objects that have collapsed at a given redshift is a direct consequence of the underlying initial gaussian random density field. Press & Schechter (1974) (PS) realized that although the small-scale modes have become nonlinear, the large scale modes still are very close to the linear prediction. By filtering the linearly evolved density field with filters of different sizes, which is equivalent to different masses ($M_{filter} \sim \rho_{mean} r_{filter}^3$), for each point in space it can be determined whether or whether not it is part of a collapsed region (i.e. having reached the linearly extrapolated density contrast threshold of ≈ 1) at this epoch. This is nothing else than the fraction of mass in the Universe in collapsed objects $F(> M)$ at that time. The number of objects of mass M that are collapsed, called the mass function can be determined by differentiation of F:

$$\frac{M\,f(M)}{\rho_{mean}} = \left|\frac{dF}{dM}\right| \qquad (1.34)$$

The PS theory uses a spherical top-hat filter and has to introduce a fudge factor of 2. The theory gives good but not perfect agreement with numerical simulations. Later careful reworkings (Bond et al. 1991) identified the origin of the fudge factor that was part of the original PS theory. The state-of-the art is the work of Sheth, Mo & Tormen (2001a,b) who computed the collapse of ellipsoidal instead of spherical regions. They fitted their result to high resolution N-body simulations. Their best fit model is:

$$\nu\,f(\nu) = 0.6444\left(1 + \frac{1}{1+\nu^{0.6}}\right)\left(\frac{\nu^2}{2\pi}\right)^{1/2}\exp\left(\frac{\nu^2}{2}\right). \qquad (1.35)$$

Here the mass variable ν is defined as

$$\nu := \frac{\delta_{ac}(z)}{\sigma(m,z)} \qquad (1.36)$$

where $\delta_{ac}(z)$ is the linearly extrapolated critical threshold for the collapse of a *spherical* region. The value for an $\Omega_m = 1$ Universe is 1.686. For the general case the values can be found in Eke, Cole & Frenk (1996). $\sigma(m,z)$ is the *rms* of the density fluctuations on scales k corresponding to the mass m. It can directly be obtained by convolving the power spectrum with a window function (see for example Peacock (1993)).

In a Universe dominated by cold DM the smallest structures collapse first. The larger ones build up by merging of the smaller ones. This formation process is therefore often termed "bottom-up".

1.2.7 Limits of the standard model

The standard model of cosmology has proven to be very successful and a large number of observables (element abundances, background radiation, matter clustering) can be described with only a handfull of free parameters.

Nevertheless some severe problems still remain. The first one is to explain the impressive homogeneity of the Universe. At the moment of the last scattering of the CMB photons, i.e. at the recombination epoch, only regions corresponding to about 1 degree on the sky were causally connected. The CMB background radiation however is homogeneous over *the whole sky*.

The second one is the fact that the Universe is flat, i.e. $\Omega = 1.02 \pm 0.02$ according to WMAP. At the time where quantum gravity ceased to be dominant, about 10^{-43} seconds after the Big Bang, the total density had to differ from one by less than 1 part in 10^{60}. Otherwise Ω would have grown to a value larger than the observed one.

An other problem is that there is no sign of anti-matter in the Universe although initially, when the temperature was higher than $T_p = m_p c^2 / k_B$ protons and antiprotons should have been present in equal amounts.

Fourth, not only the nature of the DM is completely unknown to date (see also chapter 4) but also a rough estimate of the cosmological constant Λ yields a value that is about 122 orders of magnitude too large.

Finally the nature of the expansion and the expansion mechanism are unclear (Peacock 1993).

The horizon problem and the flatness problem can be understood with the theory of inflation. The matter-antimatter problem is probably a result of broken symmetry in the very early Universe. There is hope that the theory of quantum gravity will provide solutions to the remaining questions.

1.3 Galaxy formation

In the previous sections we have discussed the current understanding of structure formation in the Universe. We will now turn to the processes relevant for galaxy formation. The model we describe here was first proposed by White & Rees (1978) and White (1978).

1. The Standard Model of Galaxy Formation

This model lives in the framework of a hierarchical build-up of structures from an initial nearly homogeneous DM and gas distribution. The gas, which became neutral at the epoch of recombination, initially has the element abundances predicted by BBN (see above).

Note that we continue to *neglect* the feedback effects of the gas on the structure formation, i.e. change of the density profile due to pressure forces, for example. The treatment described here is therefore not self-consistent. It is, however, a very good approximation, as the amount of gas is very small compared to that of the DM (see Table 1.1).

The advantage of this approach is that the model of galaxy formation can be applied to the *result* of a simulation of structure formation. This is usually very fast. A self-consistent treatment of the main physical processes can be computationally more expensive by two orders of magnitude (Yoshida et al. 2002) but has been done, too (Springel & Hernquist 2003).

1.3.1 Cooling

The homogeneously distributed neutral gas collapses together with the DM to gravitationally bound objects. During the relaxation process this gas is expected to shock heat to high temperatures and to get completely ionised ($T_{virial} > \mu m_p/(2k_B)V_{vir}^2$. μ is here the mean molecular weight for an ionised gas of the primordial composition ($\mu \approx 0.57$). This gas then will radiate its energy away because cooling is very efficient for temperatures above $T > 10^4$ K (see Figure 1.2). The resulting radiation can be observed in X-rays. The cold gas, having now less energy, can settle down in the centre of the potential well produced by the DM halo. Parts of the cloud of cold gas can collapse under gravitation and from stars.

1.3.2 Star formation

The process of star formation out of clouds of cold gas is still poorly understood. The current understanding is that the collapsing regions have sizes of 10 pc and masses of 10^5 M$_\odot$, typically. The star formation begins in dense cores of the substructure that are present in the cold clouds, often called *giant molecular clouds*.

The formation of the substructures is a subject not yet well understood by astronomers. It very probably is a combination of the influences of self-gravitation, magnetic fields, turbulence, and pressure gradients within the clouds. If a core contains more mass than the Jeans mass

$$M_{Jeans} = \left(\frac{5k_B T}{G\mu m_p}\right)^{3/2} \left(\frac{3}{4\pi\rho}\right)^{1/2}, \qquad (1.37)$$

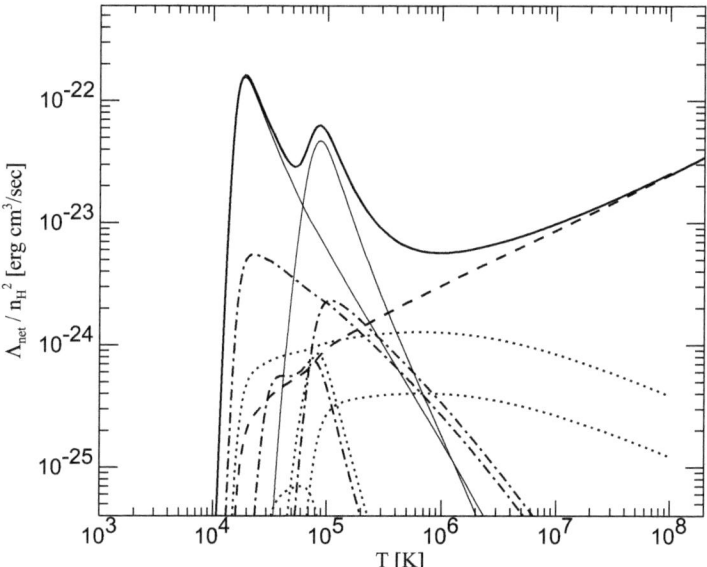

Figure 1.2: This figure shows the contributions of the line cooling efficiency of Hydrogen (thin solid lines), Helium (dot-dashed lines) as well as the one of the bremsstrahlung (dashed line) to the total cooling function (thick solid line) of an ionised plasma in primordial composition. (Plot from Volker Springel using data from Katz, Weinberg & Hernquist (1996))

1. The Standard Model of Galaxy Formation

where μ is the molecular weight of the gas and m_p the mass of the proton, the gas pressure is smaller than the self-gravitation and the cloud collapses. Once the gas is no longer transparent, radiation pressure slows down the collapse. When the temperature in the centre reaches $T_{fusion} = 10^7$ K, hydrogen burning starts and a new star is born.

For a cold gas region, the simplest model, the one which will be used in chapter 2, is that where the number of stars that form in a given time interval is proportional to the mass of cold gas available and inversely proportional to the dynamical time of the system.

The stars evolve depending on their initial masses. We use an evolution model for a whole population of stars in chapter 2 (Bruzual A. & Charlot 1993).

Whereas we neglect feedback of stars as far as radiation is concerned, the feedback from supernovae is an important ingredient in the galaxy formation model.

1.3.3 Feedback

At the end of their stellar evolution, stars with masses larger than about 4 $M_\odot$ will collapse after the pressure support from the nuclear burning has stopped and the inner region collapses to form a very high density core. Outer layers of gas falling onto this core bounce off it producing a very strong shock wave in the still infalling layers. The energy released in such a "supernova" is about 10^{51} erg $= 10^{44}$ J. This energy is released into the interstellar medium (ISM) "instantaneously" compared to the time scales of the stellar evolution.

It is uncertain however, how this energy affects the further evolution of the ISM. Current models use the global or partial reheating of the cold gas disk or the ejection of parts of the gas in "winds". In addition to the energy injection, the metallicity (we call "metals" all elements heavier than Helium) of the cold gas changes due to the production of heavy elements during the supernova explosion.

Both, the energy injection and the metallicity change, have strong effects on the cooling properties of the cold gas. The actual star formation is the result of the complicated interplay between matter accretion, gas cooling and this feedback.

1.3.4 Merging

In the hierarchical scenario of structure formation, galaxies are assembled by a combination of accretion of smooth matter and frequent merging events. The current understanding is that if the masses of two merging galaxies differ considerably, the larger system just incorporates the gas and stars from the smaller one. The disk of the larger galaxy will survive the impact. For major merging events, however, i.e. when the merg-

ing galaxies are of comparable size, the disks are supposed to be destroyed and, when the system relaxes, a large stellar bulge remains.

These models have been verified with simulations (see Springel & White (1999) and references therein). These simulations show that the disks can survive if the mass ratio of the merging galaxies is of the order of 0.3 or less.

In the hierarchical model, according to the *merger hypothesis*, disk galaxies form by smooth accretion of matter whereas bulge galaxies (i.e. elliptical galaxies) are the result of a major merging event. Between these two extremes lie galaxies that have had a major merger in the past and accreted a disk later on.

1.3.5 Reionisation

Although the Universe became neutral at recombination, observations show that both hydrogen and helium in the intergalactic medium of the local Universe are fully ionised.

The results from WMAP are consistent with a reionisation epoch of $12 < z_{reion} < 22$ at 68% confidence whereas previous estimates, especially supported by the detection of the Gunn-Peterson trough (Gunn & Peterson 1965) in the spectra of high redshift quasars, seemed to point to redshifts of $z_{reion} \approx 6$.

The sources of ionising photons can be the first stars that have formed in the Universe, special populations of very massive, bright stars (Population III stars), active galactic nuclei (AGN) and/or quasars (QSO). Simulations indicate that with reasonable assumptions, the ionising photons produced by the first stars could be sufficient to ionise the Universe (Ciardi, Stoehr & White 2003b; Ciardi, Ferrara & White 2003a).

QSO absorption lines may indicate that He$_{II}$-reionization happened around $z_{He_{II}-reion} \approx 3.3$ (Theuns & Zaroubi 2000).

1.4 Numerical simulations

The observation of the CMB and the extraction of the power spectrum of density fluctuations just 300,000 years after the Big Bang places cosmologists in the extraordinary situation of knowing - statistically - the initial conditions of the matter dominated Universe! Following a realisation of an initial density field with a given fluctuation spectrum is in principle straight-forward if star formation and radiation are neglected, as in that case only expansion and gravity have to be taken into account (Efstathiou et al. 1985). The fact that the result of such computations is *independent* of parameters - other than the cosmological ones which are accessible through observation - constitutes the power of the approach.

1. The Standard Model of Galaxy Formation

Figure 1.3: Plot taken from Ciardi, Stoehr & White (2003b) showing the propagation of reionization fronts in a ΛCDM Universe using the M-series simulation described in the next chapter.

1.4 Numerical simulations

This outcome can then be confronted with the observed Universe at low redshifts (see chapters 2, 3 and 4). In this way the initial and the final state can be observed and at the same time the intermediate evolution is well understood. Cosmological simulations therefore overcome the fact that experiments with the Universe as a whole are impossible.

In principle the evolution of the density field could be evaluated on a fixed grid, as is usual for evolution of fluid fields where pressure forces play a role. In the case of cosmological simulations however, where densities can reach values of 10^6 times the mean density the *eulerian* grid approach is inadequate. It turns out that for pure DM simulations *lagrangian* methods, i.e. the direct integration of Newton's laws for a large number of "particles" work best.

In this approach, the initial density field is represented by collisionless "particles" (typically with masses from 10^{11} to 10^5 M$_\odot$). For the initial distribution of particles, the Zel'dovich approximation (Zel'dovich 1970) of linear theory is used.

1.4.1 Zel'dovich approximation

The main idea is to use linear theory to displace the particles from their initial (unperturbed) positions according to the given density field and, in addition, to assign velocities to them. In doing so the paths of the particles can be followed instead of having to compute the evolution of a density field.

In the linear regime, the *lagrangian* position **x** of a particle being originally at its unperturbed *eulerian* position **q** is (Zel'dovich 1970)

$$\mathbf{x}(t) = \mathbf{q} - D_g(t)\,\boldsymbol{\Psi}(\mathbf{q}). \tag{1.38}$$

The *displacement field* $\boldsymbol{\Psi}(\mathbf{q})$ is related to the density contrast via

$$\delta(\mathbf{q}) = -D_g(t)\,\boldsymbol{\nabla} \circ \boldsymbol{\Psi}(\mathbf{q}) \tag{1.39}$$

The initial velocity pointing in the direction of the displacement can be obtained directly from the displacement field with

$$\dot{\mathbf{x}}(t) = -\dot{D}_g(t)\,\boldsymbol{\Psi}(\mathbf{q}). \tag{1.40}$$

The initial positions **q** of the particles must be distributed in a way that the density field they generate has itself only very small fluctuations on all scales so that they do not change the applied initial fluctuation spectrum. Two methods of generating the initial unperturbed positions are used frequently: grid and glass distributions (section 2.2.1.2).

When setting up the initial conditions of a cosmological simulation, the particles are displaced with the Zel'dovich approximation as far as possible without entering the nonlinear regime. The corresponding starting redshifts are typically of the order of $10 < z_{init} < 200$ (see also section 3.2.1.3).

1. The Standard Model of Galaxy Formation

1.4.2 Simulation codes

Although the evolution of the particle distribution is, in principle, straight-forward, performing this integration for a large number of particles (typically 10^8) is a very challenging task. The reason is that gravitation is a long range force with only positive "charge". No shielding effects render the problem local: The influence of each particle has to be computed for all others. Intrinsically this problem requires $O(N^2)$ operations for N particles. For reasonable amounts of cpu-time the direct summation can only be done for a very small number of particles.

Different techniques have been invented to circumvent this restriction. All techniques rely on an approximation of the total force acting on a given particle by approximating the contribution from relatively distant particles.

In "PM" codes, the density field produced by the particles is computed on the points of a cartesian grid and the Poisson equation is solved on the gridpoints. This can be done relatively fast in k-space using Fast Fourier Transform Techniques (FFT) resulting in the values of the gravitational potential on the grid. The force on the particles is then computed by deriving and interpolating the potential values. The computational effort of the FFT is $O(N_{grid} \log(N_{grid}))$.

As the resolution of this method is limited to the grid size, a further improvement was to compute the forces of the particles in cells around the considered particle by direct summation (P^3M, see Efstathiou & Eastwood (1981); Efstathiou et al. (1985)).

A small revolution was the invention of simulation codes using hierarchical trees (Barnes et al. 1986). This method decomposes the simulation region recursively in nested subcubes until there is only one or no particle in each of them. If ever a cell is seen from the particle for which the force shall be computed under an angle smaller than a threshold value, the centre-of-mass distance and the total cell mass are used instead of computing the sum of the forces for all the particles in the cell. If the angle is larger, the tree is followed and the procedure is repeated for all of the 8 subcubes.

With this method, as for the PM algorithm, only $O(N_{grid} \log(N_{grid}))$ operations are required. The big advantage however is, that the tree adapts automatically and without limitations to any particle distribution, i.e. there is no spatial resolution limit. The latter renders it perfect for cosmological structure formation.

We have used the treecode GADGET written by Volker Springel (Springel, Yoshida & White 2001b) for the simulations in chapter 2. This code is publicly available under the General Public License (GPL). For the simulations performed for chapters 3 and 4, we have used GADGET-2.0. In this version, Volker Springel has added a PM algorithm for the computation of the forces from very distant particles. This TreePM code is more than an order of magnitude faster than the original version.

1.4.3 Semi-analytic models

DM-only simulations have the inconvenience that by construction the simulation results cannot directly be compared with observations. Either DM-related quantities like circular velocity curves, have to be extracted from the observations or additional assumptions of the relation between mass and "light" have to be used. This is by no means obvious.

The model of galaxy formation described in sections 1.3 to 1.3.5 can, however, be used to create the missing link. Simple analytical recipes for the relevant physical processes can be coupled to the result of a structure formation computation. This *semi-analytic* approach, pioneered by White & Frenk (1991) and Kauffmann, White & Guiderdoni (1993) has been widely explored ever since (Somerville, Primack & Faber (2001);Devriendt et al. (1999); Cole et al. (2000) and references therein). The structure formation computation has to provide the mass distribution of objects at any redshift as well as the complete merging history. Several methods have been used.

White & Frenk (1991) worked with the global evolution of DM haloes as predicted by the PS-theory. Kauffmann, White & Guiderdoni (1993) followed merging, cooling, starformation and feedback in merging-trees they generated with Monte Carlo techniques using the PS mass functions and the clustering model from Mo & White (1996). The advantage of these methods is, that only rather limited computational resources are required.

Other approaches combine the semi-analytic galaxy formation model with the result of a cosmological DM-simulation. Kauffmann et al. (1997) use the positions of the DM-haloes at $z = 0$ and their previous method to generate merging-trees. Kauffmann et al. (1999) take the merging-trees directly from the N-body simulation. Springel et al. (2001a) finally extend this method by including merging-trees for substructure haloes.

There are several advantages to the approach where semi-analytic techniques are combined with N-body simulations. The first one is that the positions and the velocities of the DM-haloes, and therefore also of the attached galaxies, are available. This makes it possible to study galaxy correlations, to look for environmental effects and to produce mock catalogues. The second one is that the merging history is correct and not only statistically reproduced. All possible dependencies of the assembly of the DM-haloes on the local density, tidal torques and nearby objects are fully accounted for in the galaxy catalogues. We will use this second method in chapter 2.

The application of the semi-analytic model of galaxy formation to the DM-simulations multiplies the possibilities of comparison with observations. Because the underlying DM structure seems to be well understood, this coupled method allow us to test the model of galaxy formation and the corresponding physical assumptions in a very complete way.

1. The Standard Model of Galaxy Formation

1.4.4 DM-only vs. SPH

Instead of neglecting the baryonic component in the cosmological simulations, the corresponding physics can also be included in a self-consistent way. The hydrodynamic part is often implemented as smoothed particle hydrodynamics (SPH). In this method, gas "particles" carry gas properties like temperature, composition and eventually stars. The hydrodynamic equations are then solved using quantities averaged over a given number of neighbouring particles, typically 32.

Although, of course, solving the full hydrodynamics should eventually be the most accurate method, as for now the approach has several drawbacks. The resolution is set by the smoothed quantities which by construction is much poorer than the DM resolution. The computation time at given resolution is about a factor of 100 larger compared to the semi-analytic approach (Yoshida et al. 2002). In addition, for each set of parameters or recipes for the gas physics, a new simulation has to be run. For this reason only a fiducial run can be done in most cases. Parameter studies, feasible with the semi-analytic approach, are impossible. Finally the gas physics is very complicated and the simplifications may not be valid. In the semi-analytical approach, this ignorance can be much more easily studied. Nevertheless, full cosmological hydrodynamic simulations have been carried out with some success (Springel & Hernquist 2003).

For simplified cases the SPH method and the semi-analytic approach give very similar fractions of cold and hot gas for the collapsed objects when state-of-the-art SPH implementations are used (Yoshida et al. 2002).

In recent years the computation of the hydrodynamic equations on a grid has also become popular in cosmology. This is due to the advent of adaptive mesh refinement (AMR) or multigrid techniques where subgrids are nested in larger grid cells wherever the higher resolution is required.

Although SPH and AMR techniques most probably will become more and more important in the future, the semi-analytic approach is the most efficient means for modelling the formation of large galaxy populations today.

2 Galaxies and Environment

2. Galaxies and Environment

2.1 Introduction

The fact that galaxy properties depend on the local environment is established since the work of Dressler (1980). This is most clearly seen in the difference in the relative abundance of spiral and elliptical galaxies in cluster and field environments (section 1.1). It is natural to try and use simulations of galaxy formation to understand how these correlations have emerged aiming for a deeper understanding of galaxy formation as a whole.

The hybrid method mentioned in the introduction turns out to be fast and to give very similar results to a full SPH treatment of the gas physics (Yoshida et al. 2002). We will use this approach for the work in this chapter. Tackling the environmental dependence of the properties of DM haloes and of the galaxies they host requires high mass resolution in order to resolve dwarf galaxies as well as a large simulation volume in order to obtain meaningful statistics.

Given that a relatively small region must be simulated in order to obtain high resolution on individual systems, the usual choice of periodic boundary conditions in a cubic box causes problems because of the severe constraints on the Fourier content of the perturbation field which it implies.

In this situation the method of choice is the resimulation technique presented in section 2.2.1. It allows to take the large scale tidal field of the cosmological environment into account while concentrating most of the computational effort on a small previously selected region. It has already been successfully applied to clusters and groups of galaxies (Navarro, Frenk & White (1995b, 1997); Tormen, Bouchet & White (1997);Tormen (1998); Moore et al. (1999); Klypin et al. (1999); Jing & Suto (1999); Jing (1999); Ghigna et al. (2000); Springel et al. (2001a); Bullock et al. (2001); Lanzoni & Ciotti (2003) to galaxies (Power et al. (2003); Moore et al. (1999); Steinmetz (1999); Klypin et al. (1999) and also chapter two) and even to underdense regions (Stoehr 1999). However until now only one other mean density region has been studied in the full cosmological context (Mathis et al. 2002). The resimulation technique makes it possible to select a relatively small region - in our case a sphere of 52 Mpc/h in diameter - and run it down to $z = 0$. We carefully selected this sphere from the parent simulation to make sure that it is as representative as possible of the whole Universe. This ensures that quantities like the mass distribution of the DM haloes match the global expectation even at high mass where the number of objects is small.

We have performed a series of simulations of the same region with different mass resolution and have evaluated for each the properties of the DM haloes as well as the properties of the galaxies produced by our semi-analytic recipes of galaxy formation (sections 1.3 and 2.5). We explicitly study the correlations of these properties with the

mass of the corresponding DM haloes and with their local environment.

After describing the set-up and the execution of the simulation series - which we call the M-series - we present two methods for the density estimation of the environment of a DM halo (section 2.3). We then show the results of the DM properties of the haloes in our simulations (section 2.4). In section 2.5 we describe and apply the semi-analytic galaxy formation model to the simulations and present our findings.

2.2 N-body simulations

2.2.1 Initial conditions

The initial conditions for the simulation series were prepared using the Zoomed Initial Conditions code ZIC written by Bepi Tormen (Tormen, Bouchet & White (1997), see also Stoehr (1999)). This code package allows the user to select particles from a previously computed parent simulation, to track them back to their initial (lagrangian) positions and to determine a corresponding connected lagrangian region to be followed at high resolution. This region can be of arbitrary shape. It is often convenient to render it convex and to add a boundary shell in order to prevent "contamination", i.e. LR particles falling into the region of interest.

This high resolution region (HR region) is then placed at the centre of the new re-simulation volume. The particles of the initial conditions of the parent simulation are regrouped on a spherical grid of given angle θ where the radial cell bounds are chosen so that the resulting cells are approximately "cubic". This results in low resolution particles (LR particles) with radially increasing masses. The angle θ or alternatively the number of particles representing the structure of the parent simulation controls the accuracy of the corresponding tidal field. We used a $2°$ grid, considerably smaller than the $3°$ grid considered necessary by Stoehr (1999). The distribution of the LR particles was the same for all simulations of the M-series.

As the HR region is centred in the new simulation box, it is feasible to remove the outmost LR particles in order to obtain a simulation sphere instead of a simulation box. This allows the resimulation to be computed with vacuum boundary conditions. This is advantageous because the computation of the latter can consume up to 20% of the cpu-time in a periodic simulation.

LR particles falling in the previously determined HR region are removed and the HR region is filled with a homogenous distribution of HR particles of the desired mass resolution. These particles have to be displaced from their positions according to the displacement field of the parent simulation as well as with a newly generated displacement field corresponding to density fluctuations of higher spatial frequency than those

2. Galaxies and Environment

present in the parent simulation (see section 1.4.1).

One advantage of the method implemented in ZIC is that the HR region can be chosen with arbitrary shape. This allows the computational effort to be concentrated on the region of interest, i.e. the one that will at $z = 0$ contain the resimulated object. If instead the smallest cubic HR region is selected, typically three times or more HR particles may be necessary, significantly reducing the mass resolution that can be achieved with a given amount of cpu time.

A second important advantage of this code is that it is possible to extract the displacement field from an existing simulation and thus carry out resimulations from initial conditions that have been produced with a different code.

In order to assure no contamination, even with the arbitrary shaped lagrangian HR region, it is necessary to follow more HR particles than will later form the resimulated object. As a rule of thumb, for resimulations of galaxy clusters about three times more HR particles are necessary than are contained within r_{200} of the final object. For a resimulated galaxy this increases to a factor of four or five. For the mean density region considered in this chapter only about two times more HR particles were required. The best situation however is found when resimulating voids. In this case nearly no HR particles other than the ones found in the void have to be set up.

Several similar codes for setting up resimulations exist. The most prominent are the ones implemented by Adrian Jenkins (Durham, UK) and Edmund Bertschinger (MIT). Whereas the first implementation is very similar to the method used by ZIC, the latter uses cartesian grids only and has implemented a convolution method to generate the required displacement fields (Bertschinger 2001).

The advantage of resimulations using cartesian grids only is that when replacing a LR grid cell with a HR grid, even the density at the interface is close to be homogenous. This is not true for the spherical grids where the inclusion or omission of LR particles very close to the border of the HR region necessarily produces small over- or underdensities. This favours mixing of LR and HR particles in the boundary region leading to slightly larger contamination.

2.2.1.1 Region selection

The parent simulation we used was the 512^3 ΛCDM simulation set up and run by Naoki Yoshida (Yoshida, Sheth & Diaferio 2001) who also adapted the GADGET reading routine to the ZIC output and produced the data necessary for the inclusion of the parent displacement field. In addition to this and making available the simulation data he also kindly provided a halo catalogue which was used for the selection of the spherical region to resimulated.

2.2 N-body simulations

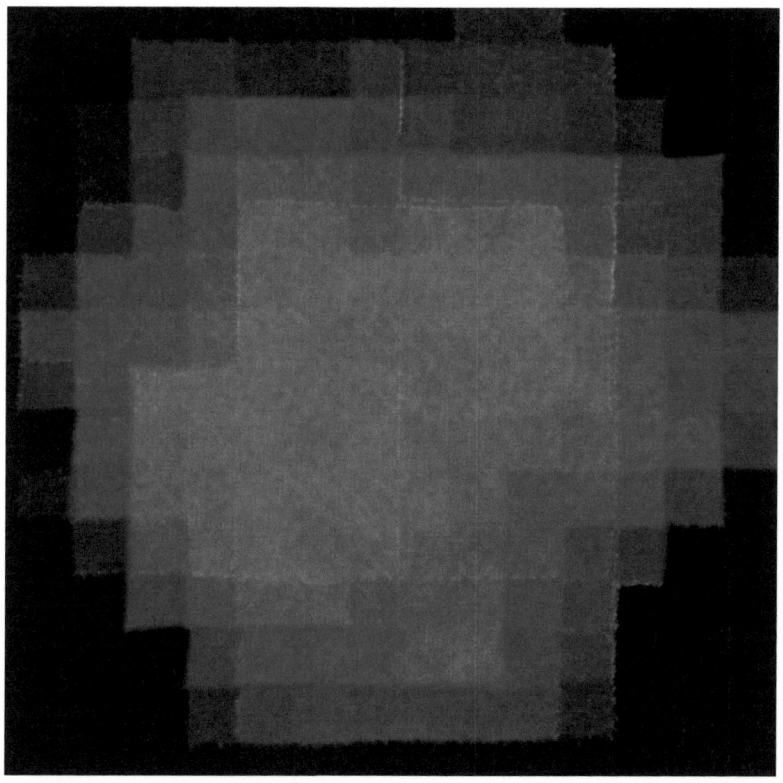

Figure 2.1: HR region of M3 at redshift $z = 20$. The shape was chosen so that at $z = 0$ the resimulated sphere contains no LR particle. The cubic region shown has a comoving side length of 70 Mpc/h.

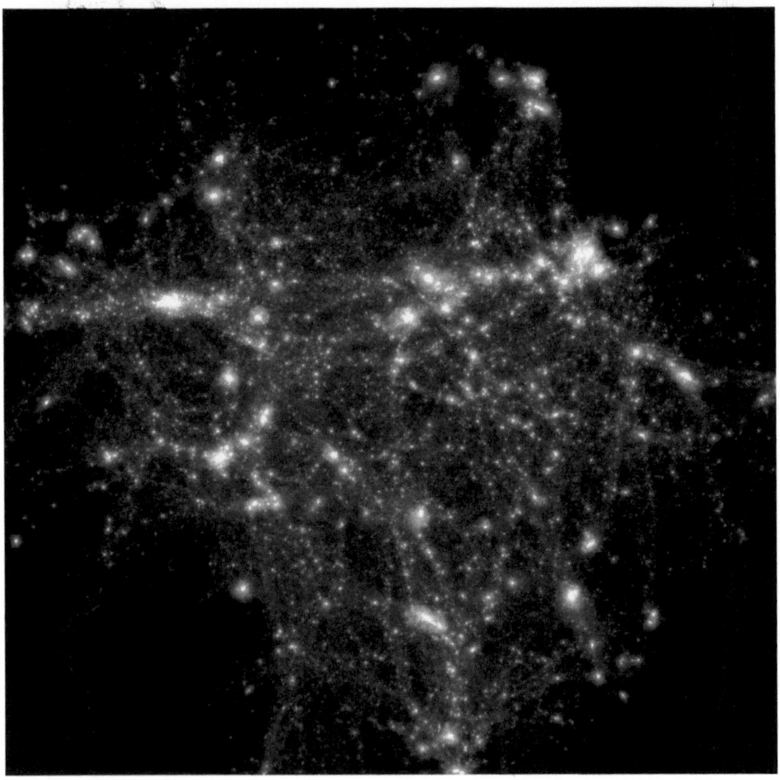

Figure 2.2: As Figure 2.1 but for the matter distribution at $z = 0$.

2.2 N-body simulations

This region was determined by optimising the mean values of the number of haloes, the mass in haloes as well as a requiring the net matter flow through the surface of the selected sphere to be as close to zero as possible. Whereas the first two conditions fix the mass function, the third one assures slow evolution of the boundary close to $z = 0$.

For the desired mass resolution of about 1.6×10^8 $M_\odot/h$ and given our maximal particle number of 7×10^7, the largest uncontaminated sphere to resimulate had a radius of 27 Mpc/h. We selected randomly 20000 such spheres from the parent simulation and selected the one with the smallest deviation from our criteria above.

After a HR boundary layer had been added, we ran the lowest resolution simulation of the series at the mass resolution of the parent simulation. We then identified LR particles that fell into the final sphere, added the corresponding regions of the initial conditions to the HR region and reran the simulations. The largest sphere containing no LR particles has in the parent simulation the coordinates:

$$
\begin{aligned}
x_{center,P} &= 344.653 \text{ Mpc/h} \\
y_{center,P} &= 240.128 \text{ Mpc/h} \\
z_{center,P} &= 298.280 \text{ Mpc/h} \\
radius &= 26 \text{ Mpc/h}
\end{aligned}
\tag{2.1}
$$

This corresponds to

$$
\begin{aligned}
x_{center,M} &= 4.1497 \text{ Mpc/h} \\
y_{center,M} &= -5.9124 \text{ Mpc/h} \\
z_{center,M} &= 0.38836 \text{ Mpc/h}
\end{aligned}
\tag{2.2}
$$

in the coordinate system of the resimulations where the centre of mass of the eulerian HR particles has been put at the simulation centre.

2.2.1.2 Grid and glass distributions

Two types of unperturbed HR particle distributions are commonly used. The most popular, because most easily implemented, distribution is to lay out the HR particles on a cartesian grid. The advantage is that this distribution is perfectly uniform. The inconvenience is however, that the grid introduces very strongly preferred directions along its axes.

As second possibility is the use of a "glass" distribution of particles. Such a particle distribution is obtained by evolving a set of randomly distributed particles with repulsive

"gravitation" forces until their positions do not change any more (White 1993). This method overcomes the problem of the cartesian-grid distributions. If the glass distribution is evolved long enough, the density fluctuations of the unperturbed distribution can be very small, more than sufficient for the use in the cosmological simulations.

The homogeneity of the particle distribution of glass files allows arbitrary regions to be extracted where the density in the region boundary is always close to the mean density. Cutting out regions from grid distributions has to be done carefully: the cut has to be made exactly between two gridpoints in order to conserve homogeneity.

Although a glass distribution appears more random than a grid distribution, the distribution of the separations between neighbouring particles is very narrowly peaked around the mean interparticle separation.

2.2.1.3 Simulations

The cosmological parameters of the parent simulation and therefore also of the series of resimulations are given in Table 2.1.

Ω_m	0.3
Ω_Λ	0.7
Ω_c	0.0
σ_8	0.9
Γ	0.21
n	1.0
h	0.7
z_{parent}	35

Table 2.1: Cosmological parameters used for the M-series simulations.

The transfer function of Bond & Efstathiou (1987) (equation 1.17) was used to generate the density fluctuations. The corresponding powerspectrum is shown in Figure 2.3 (dashed) together with the powerspectrum as computed by CMBFAST (Seljak & Zaldarriaga 1996) (solid).

We followed Springel et al. (2001a) in the choice of the mass resolution. The parameters of the resimulations are summarised in Table 2.2.

All simulations were ran on the Cray T3E of the Rechenzentrum of the Max-Planck Society in Garching. Depending on the starting redshift, the LR particles are displaced

2.2 N-body simulations

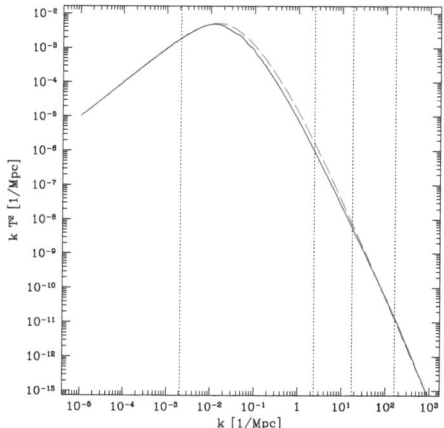

Figure 2.3: Powerspectra from Bond & Efstathiou (1987) (dashed) and from CMBFAST (solid). The vertical lines indicate for M3 (from left to right) the largest fluctuation mode of the parent simulation, the matching frequency of the LR and HR particles, and the Nyquist-frequency of the HR particles. The rightmost vertical line shows the Nyquist frequency of the SHR particles of GA3n (see section 3.2.1.2)

2. Galaxies and Environment

	m $[M_\odot/h]$	nhr	nlr	$ntot$	z_{ini}	ϵ [kpc/h]	cpu [proc. hr]
M0	6.827×10^{10}	168436	602272	770708	35	15	121
M1	4.814×10^{9}	2388896	602272	2991168	35	6	730
M2	9.529×10^{8}	12067979	602302	12670281	70	3	3300
M3	1.668×10^{8}	68958296	602341	69560637	120	1.4	67000

Table 2.2: Mass resolution m, number of HR particles nhr, number of LR particles nlr, total particle number $ntot$, starting redshift z_{ini}, gravitational softening ϵ and CPU times for the four M-series simulations.

more (M0) or less (M3) from their unperturbed positions which leads to the inclusion or exclusion of LR particles that are very close to the HR region. This is the reason for the small differences in the LR particle numbers.

In order to be sure that the particle representation of the density field behaves like a collisionless fluid, two-body effects have to be suppressed. This is done by modifying the interacting potential on small scales so that a constant finite value is reached if two particles have identical positions. The parameter ϵ – the gravitational softening – given in Table 2.2 corresponds to a finite particle size and thus to the formal resolution of the simulation. The softening was kept fixed in comoving coordinates down to $z = 6$. Below $z = 6$ it was kept fixed in physical coordinates. The effective spatial resolution limit is problem dependent can only be determined by convergence studies. It is approximately about three to four times the gravitational softening (section 4.3).

The effective mass resolution, e.g. for convergence in the abundance of haloes of a given mass, is discussed in section 2.4.1. Note that convergence studies are often the only way to assess the reliability of simulations.

During the analysis of the simulations, some integration inaccuracies of the simulation series turned up that might be of interest for future users of the M-series data. The masses of the HR particles differ by a factor of $h = 0.7$ from the corresponding S-series simulations (Springel et al. 2001a). Secondly the different resolution simulations do not have the same phases for their small scale modes. Structures influenced by modes with $k > 2.35$ Mpc^{-1} differ from simulation to simulation. Note that this is not true

Random seed	12345
LR mesh size	512
HR mesh sizes	24/52/88/156
HR replication	10
HR/LR matching frequency	2.35 [1/Mpc]
LR grid resolution	2.0 [deg]

Table 2.3: Parameters used for the generation of the initial conditions. The HR mesh sizes are given for the simulations M0, M1, M2 and M3, respectively.

for the GA-series simulations used in chapter 3 and chapter 4. Due to an incorrect interpolation routine in ZIC, small deviations from the correct displacement appeared at the boundaries of the 512^3 parent grid cells. At very high redshifts a corresponding structure can be seen in images of thin slices through the particle distribution. This problem was corrected for the simulations in the following chapters.

As mentioned before, when placing the HR region into a grid particle distribution, over- or underdense regions can be produced at the region boundaries. A small underdense region has developed in the M-series simulations just at the border of the uncontaminated region (at about (-20 Mpc/h,-20 Mpc/h, 0 Mpc/h). As the HR region for the simulations of the upcoming chapters was a glass distribution, the density at the HR region boundaries was perfect.

Finally, the simulation series was run with the code GADGET in its version 1.0 and with the time-step criterion "1". Extensive tests showed that this selection can lead to densities in the very inner parts of the DM haloes that are too high.

For the analysis in this work however, none of these inaccuracies is relevant.

2.3 Halo environment

The Press-Schechter formalism (Press & Schechter 1974; Bond et al. 1991; Sheth et al. 2001b) allows to compute the mass function of collapsed DM haloes at any redshift to be computed from the initial power spectrum, i.e. for a given set of cosmological parameters. In addition, the formalism allows to derive the average progenitor mass function of the DM haloes and thus to follow some aspects of their accretion history. This can be studied directly or can be the basis of semi-analytic models of galaxy formation (Cole

2. Galaxies and Environment

et al. (2000), see also section 1.4.3).

Analysing the GIF simulations (Kauffmann et al. 1999), Lemson & Kauffmann (1999) studied the correlations of halo properties like the mass, the spin parameter and the formation time with the halo environment. Besides the obvious correlation between the halo mass and the density of the local surroundings they found a weak correlation of the spin parameter with environment. We confirm these results below.

2.3.1 Estimators

For the estimation of the local density of a DM halo, a characteristic length scale has to be specified. Although in principle this choice, especially for a near-scale invariant region of the power spectrum is somewhat arbitrary, it should reflect the size of the region that may dominate local evolution if we are to be sensitive to a link between the environment and the object properties. We consider two estimators describing the local density of a DM halo and choose the characteristic scale to be 3 Mpc/h corresponding to a mass of 9.4×10^{12} $M_\odot$/h.

This choice reflects the higher mass resolution of M3 compared to the GIF simulation, where 5 Mpc/h was chosen (Lemson & Kauffmann 1999).

2.3.1.1 Mass in shells

Lemson & Kauffmann (1999) suggested to evaluate the density on a spherical shell of inner radius 2 Mpc/h and outer radius 5 Mpc/h centred on the halo position. The inner bound of the shell assures that none of the particles within the virial radius of the halo under consideration is taken into account: the virial radius of the largest halo in M3 is about 700 kpc/h. Figure 2.4 shows the halo-density distribution as a function of the halo mass for M3 as well as the histogram of the density values. Only haloes with more than 100 particles have been included.

2.3.1.2 FOF subtraction

We propose a second method. Instead of using all the mass for the density evaluation we only select regions that are not part of collapsed haloes: After the simulation particles are assigned on a cartesian grid, all grid cells containing particles that are part of collapsed objects, i.e. particles that have been selected by a standard FOF algorithm (Davis et al. 1985), are removed. This removes most of the mass from the grid. We then compute the average density of the remaining cells in a sphere with radius 3 Mpc/h. An inner cut-off is not necessary because the halo's own contribution has been taken out already. We fixed the cartesian grid cells to have side length equal to the mean interparticle

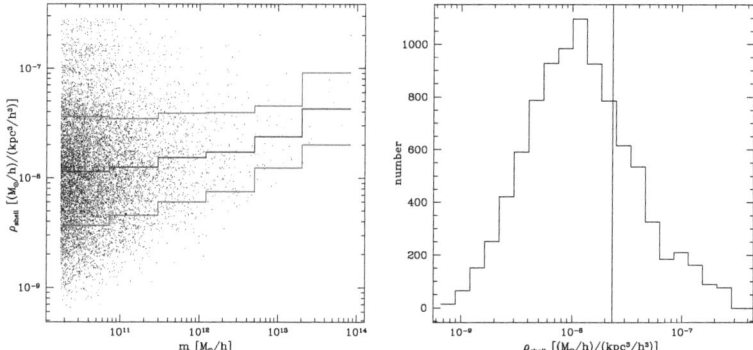

Figure 2.4: *Left panel*: Densities evaluated in spherical shells around the haloes in M3 as a function of their mass. The average values of $\log(\rho)$ and the corresponding 1σ deviations are overplotted with lines. *Right panel*: distribution of the density values. The vertical line corresponds to the mean. Only haloes with more than 100 particles have been taken into account.

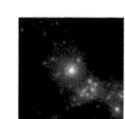

2. Galaxies and Environment

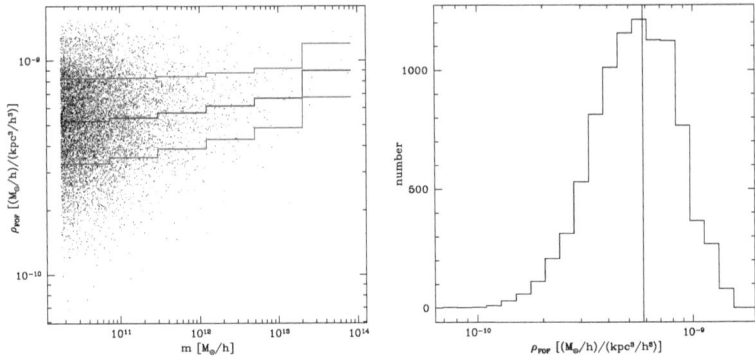

Figure 2.5: As Figure 2.4 but this time using the "FOF subtraction" method. The lines show the mean and deviation of the log-values.

separation of the HR particles. The lower panel of Figure 2.5 shows the corresponding density distribution.

2.3.1.3 Comparison

Figure 2.6 shows the correlation of the two methods on a halo-by-halo basis. Due to the different philosophy of the density estimation there is significant amount of scatter. This is an advantage as it allows to check for the robustness of the correlations found in the following subsections.

The advantage of the FOF subtraction method is that the dominance of the collapsed objects is reduced as no region with overdensities of more than 125 times the mean density are taken into account (Stoehr 1999, page 45). An advantage of the shell method is that it is easy to implement and that the evaluated densities lie around the mean density and that they are independent of the mass resolution. In addition, the shell method returns density values with a dynamic range two orders of magnitude larger.

2.4 Dark Matter Properties

Radial velocity curves of DM haloes show a clear boundary between the infalling material and the part of the halo that is virialised Lacey & Cole (1993). Several methods

2.4 Dark Matter Properties

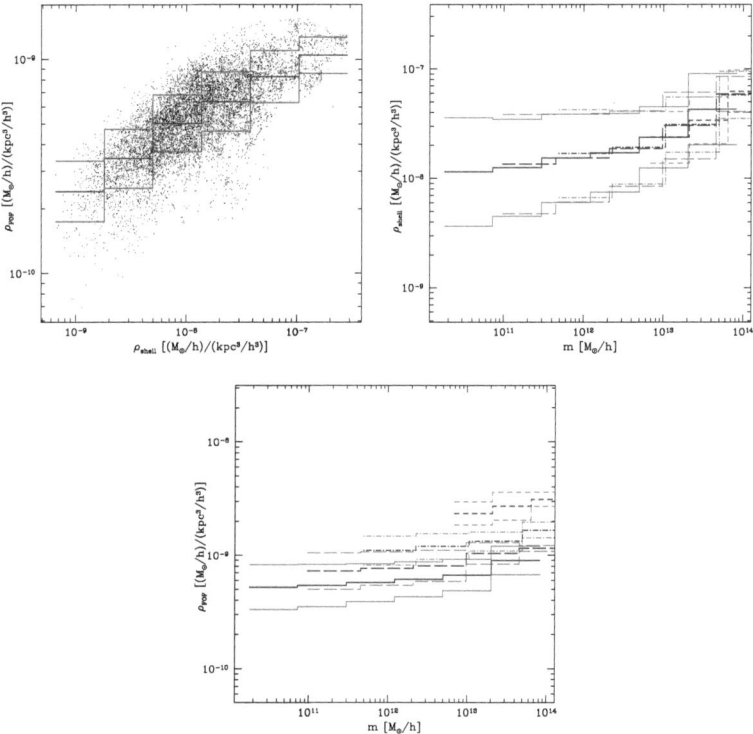

Figure 2.6: *Upper left panel:* Comparison of the two density estimation methods on a halo-by-halo basis. *Upper right panel:* Histogram of the correlations between the shell estimators and the halo mass for the four different simulations M0 (short-dashed), M1 (dot-dashed), M2 (long-dashed) and M3 (solid). *Lower panel:* As the centre panel but for the "FOF subtraction estimator". This estimator shows a rather pronounced resolution dependence.

2. Galaxies and Environment

exist to find the simulation particles corresponding to the virialised regions. The most popular one is the FOF method that links together all particles that are separated by less than a given fraction of the mean interparticle separation, typically chosen to be 0.2 following the analysis of Cole & Lacey (1996). The advantage of this method is that it is very robust and very fast. The selected objects correspond to material enclosed in an isodensity contour of $1/0.2^3$ times the mean density (see for example Stoehr (1999)). They are usually rather elongated.

A second method is to define the haloes as spheres around the most bound particle enclosing some given multiple of a characteristic density. For the work in this chapter we concentrate on the use of M_{200} and r_{200} defined as the mass and the radius corresponding to a sphere with 200 times the critical density of the Universe at $z = 0$. This (M_{200}, r_{200}) definition is the one used by Navarro, Frenk & White (1997) (hereafter NFW).

2.4.1 Mass function

Figure 2.7 shows the cumulative number of haloes with more than 20 simulation particles in the uncontaminated HR region of the simulations M0, M1, M2 and M3, respectively. The agreement down to about 50 particle haloes is very good. Near their low mass limits the mass functions are always slightly higher than their higher resolution counterparts. The reason for this is stochastic noise in the detection of small haloes. This is a well known effect of the FOF group finder. We explicitly checked in test simulations that it is not due to the timestep criterion used for the computation.

Figure 2.8 shows the mass function, i.e. the number of DM haloes per log mass interval and per unit simulation volume for different redshifts. Overplotted is the analytical fit given by Sheth et al. (2001b). Again the agreement down to about 50 particle haloes is good. At high redshifts the mass functions of the M-series are steeper than the analytical fit.

2.4.2 Profiles

Figure 2.9 shows profiles of the most massive of the DM haloes in M3. This halo with mass of 8×10^{13} corresponds to a galaxy cluster. Shown are the density profile, the rotation velocity curve

$$V_{circ}(r) = \sqrt{\frac{G\ M(<r)}{r}} \quad (2.3)$$

as well as the average radial velocity. The density profile and the circular velocity curve are well fit by a NFW profile with concentration $c = 8$. For a detailed discussion of density profiles see section 3.2.2.1 and section 3.2.2.4.

2.4 Dark Matter Properties

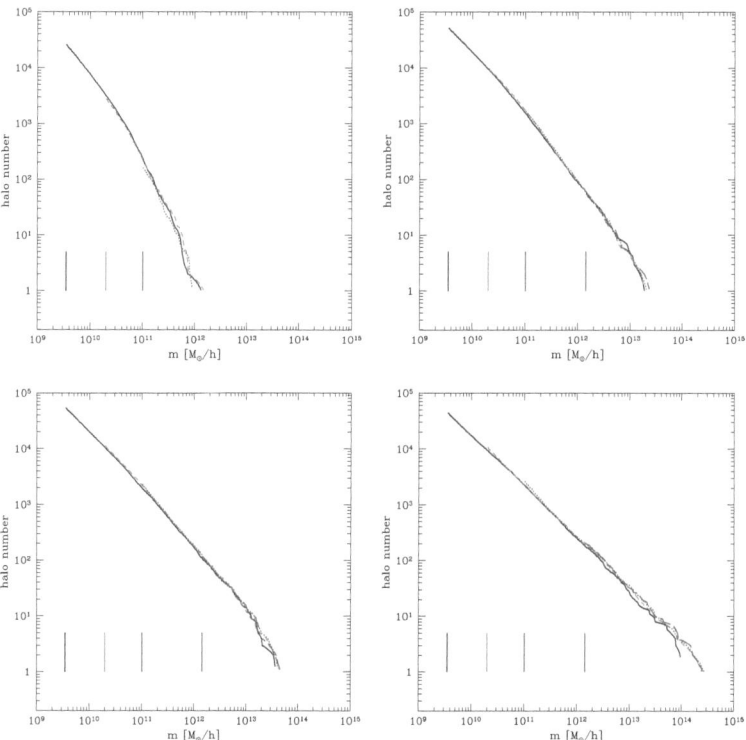

Figure 2.7: Cumulative halo numbers for haloes with more than 20 particles in the four M-series simulations at the redshifts $z = 5$, $z = 2$, $z = 1$ and $z = 0$, respectively. The vertical lines show the 20 particle limits. The masses M_{200} have been used.

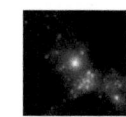

2. Galaxies and Environment

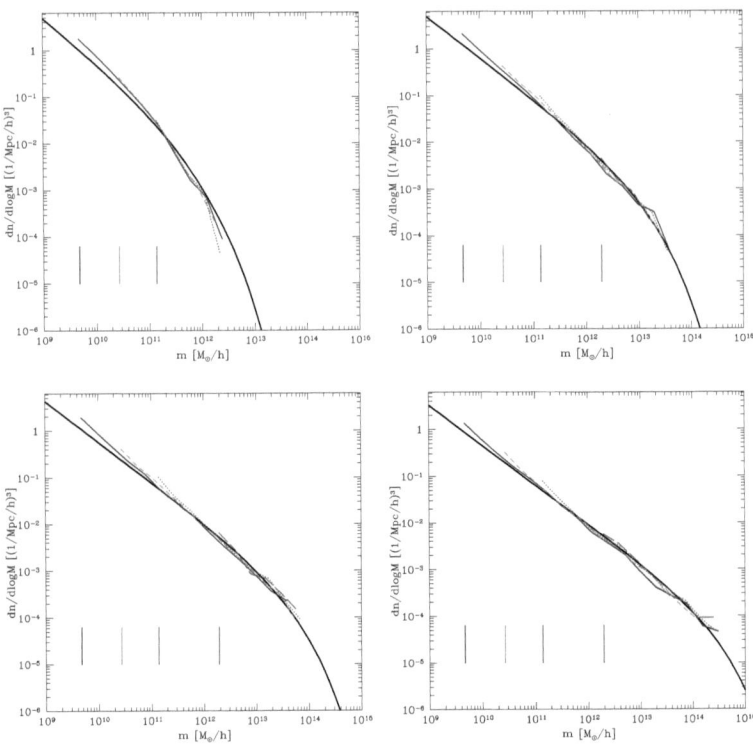

Figure 2.8: Mass functions for haloes with more than 20 particles in the four M-series simulations at the redshifts $z = 5$, $z = 2$, $z = 1$ and $z = 0$, respectively. The vertical lines show the 20 particle limits. The solid thick line corresponds to the analytic formula of Sheth et al. (2001b).

2.4 Dark Matter Properties

Due to the version of GADGET that was used for the simulations as well as the timestep criterion, the very inner part of the simulated haloes was not correctly evolved. This is shown in Figure 2.11 where the profiles of the 50 most massive haloes of M3 have been stacked together. The average rotation curve of these haloes lies well above the NFW curve for radii smaller than about $0.01 r_{200}$. We use this value to define the spatial resolution of this simulation series.

NFW profiles are determined by two parameters, for example the virial radius r_{200} and the concentration parameter c. We use only haloes with more than 500 particles, to be sure that the concentrations can be safely determined. Figure 2.10 shows the distribution of the concentration parameter as well as its correlation with mass and environment. We find the well known anti-correlation with mass (see for example NFW). If we reject the left-most bin of the "FOF"-histogram because of small number statistics (10 haloes only) we find a weak positive correlation of c with the environment with both methods.

2.4.3 Shapes

It is a long established fact that the FOF-DM haloes are not spherical but are ellipsoidally shaped with typical axis ratios for the mass distribution of typically 1:1.4:2 (Frenk et al. 1988; Dubinski & Carlberg 1991). Here we use the M_{200} definition of the DM haloes when computation of axis ratios. After determining the centre of mass of the halo particles, the inertia tensor

$$\Theta_{ik} = \sum_{\nu=1}^{N} m_\nu \left(r_\nu^2 \, \delta_{ik} - x_i^\nu \, x_k^\nu \right) \tag{2.4}$$

and its eigenvalues are computed. The halo-axis-ratios are defined as the square-root of the ratios of the lengths of the eigen vectors. The results are shown in Figures 2.12 and 2.13.

We find that the short-to-long and the intermediate-to-long axis ratio anticorrelate weakly with mass, implying that more massive haloes tend to have slightly less spherical mass distributions than less massive haloes. Our results are consistent with the fitting formula given by Jing & Suto (2002).

The average halo ratio for the particles within r_{200} is 1:1.2:1.3 and for the FOF particles about 1:1.3:1.6. Our haloes are thus a bit more spherical than those of Dubinski & Carlberg (1991). A reason for this might be that their simulations were carried out in a cosmology with vanishing cosmological constant.

We find no correlation of the axis-ratios with environment. The small trend visible is a result of the correlation of the axis-ratios with halo mass and the correlation of halo mass with the local environment.

2. Galaxies and Environment

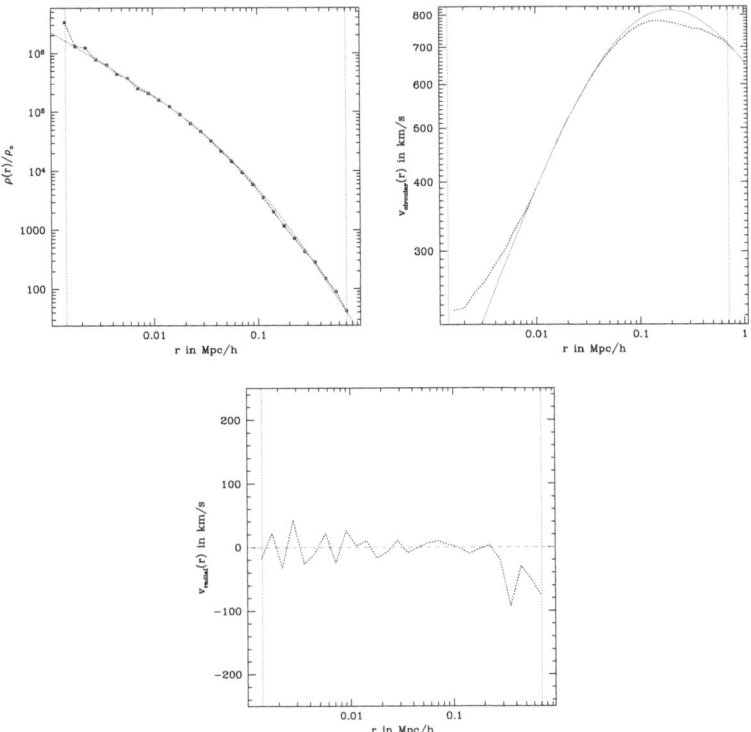

Figure 2.9: Density profile (upper left panel), rotation velocity curve (upper right panel) and radial velocity profile (lower panel) for the most massive DM halo in M3. The best fitting NFW profiles are overplotted as well the softening length (left vertical line) and the virial radius (right vertical line).

2.4 Dark Matter Properties

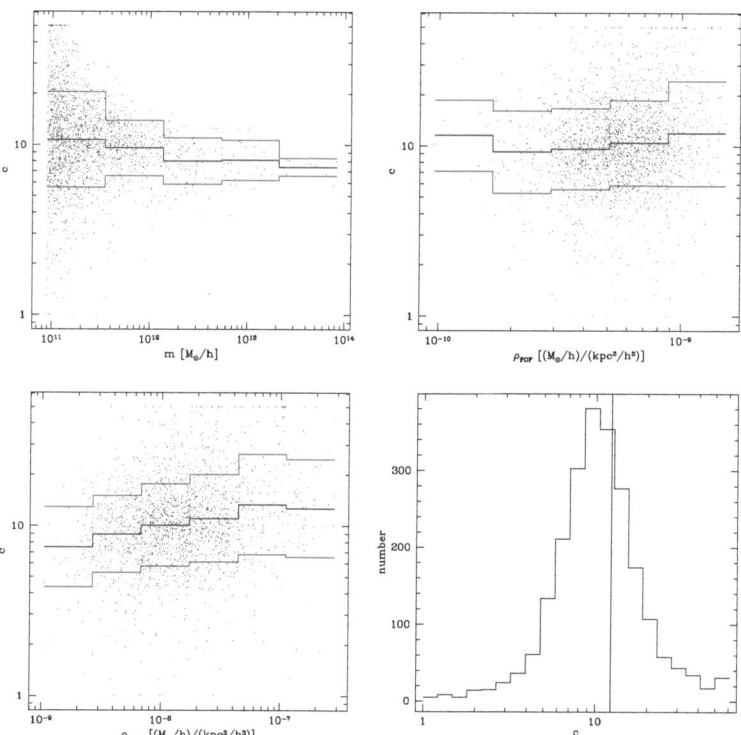

Figure 2.10: Correlation of the NFW concentration parameter c with mass (upper left) and with environment (upper right : "FOF"-method, lower left, lower left: shell method) for haloes in M3. The lower right panel shows the concentration distribution. For a small number of low-mass haloes no NFW profile fit could be found (upper right panel, points in the upper left corner).

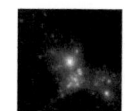

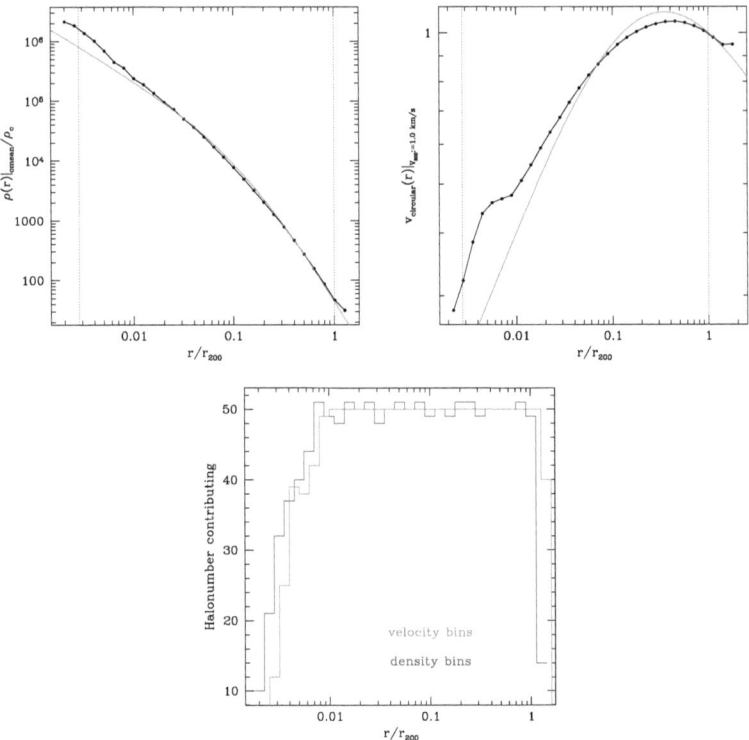

Figure 2.11: Averaged density profile (upper left panel), averaged circular velocity curve (upper right panel) and total number of haloes contributing to each bin (lower panel) for the 50 most massive DM haloes in M3. NFW profiles for the average concentration $c_{average} = 6.2$ are overplotted. Note the sharp rise of the profiles around 1% of r_{200} which is due to integration problems.

2.4 Dark Matter Properties

Figure 2.14 shows the short-to-long axis ratio as a function of the local matter density evaluated with the shell-method. We split the halo masses in mass bins from 10^{10} M$_\odot$/h to 10^{11} M$_\odot$/h (upper left), 10^{11} M$_\odot$/h to 10^{12} M$_\odot$/h (upper right), 10^{12} M$_\odot$/h to 10^{13} M$_\odot$/h (lower left) and 10^{13} M$_\odot$/h to 10^{14} M$_\odot$/h (lower right). By splitting the haloes according to their mass we reduce the influence of the mass-density correlation. The remaining correlation between density and the halo-axis ratio is consistent with zero.

2.4.4 Spin parameter

The spin parameter λ is defined as the ratio

$$\lambda = \frac{L\sqrt{E}}{GM^{2.5}}. \tag{2.5}$$

It is a dimensionless measure for the angular momentum a halo has gained from tidal torques during its evolution (Hoyle 1949; Peebles 1969; White 1984; Steinmetz & Bartelmann 1995). A rotationally supported system has $\lambda \simeq 0.4$. This parameter is traditionally computed from the FOF halo particles and was found to have a mean value of $\lambda \approx 0.05$. The distribution as well as the correlations with mass and density for both environment definitions are shown in Figure 2.15. We find a mean value of $\lambda = 0.062$ for haloes with more than 500 particles. As Lemson & Kauffmann (1999) we find a small positive correlation of the spin parameter λ with environment with both density estimators when we again reject the first bin of the histogram of the "FOF"-method.

The anticorrelation of the spin parameter with mass, shown in the upper left panel of Figure 2.15 is well known (see for example Peacock (1993)). It is due to the fact that more extreme overdensities in the initial density fields started to collapse earlier when the tidal torques on the collapsing regions were smaller.

2.4.5 Formation time

The DM halo properties analysed so far were taken from a snapshot at $z = 0$ of the halo evolution. In this and the following section we study properties of the evolution itself. The first property is the formation time, which is defined as the redshift at which a DM halo first contained half of its present mass. This time is a measure of the recent accretion rate.

Figure 2.16 shows the distribution and correlations of the formation times of haloes with more than 500 particles in M3. Again, rejecting the first bin of the "FOF-method"-histogram we find a correlation of the formation time with mass and density. This time, the correlation with the density is negative and so we check in Figure 2.17 that this is not just a result from the correlation of environment with mass. We split the haloes in

2. Galaxies and Environment

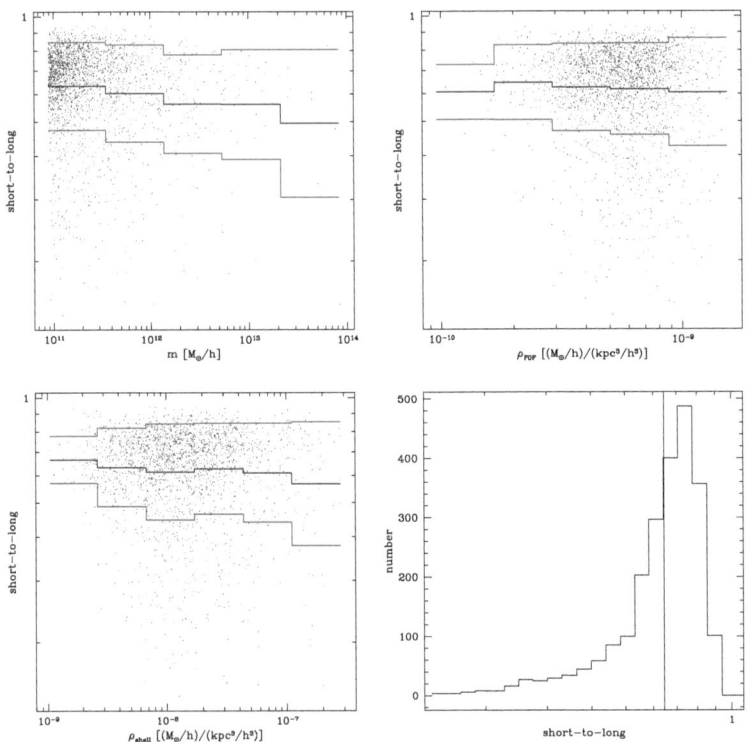

Figure 2.12: As the previous figures but for the short-to-long axis-rations of the DM haloes. All particles within the virial radius r_{200} have been used.

2.4 Dark Matter Properties

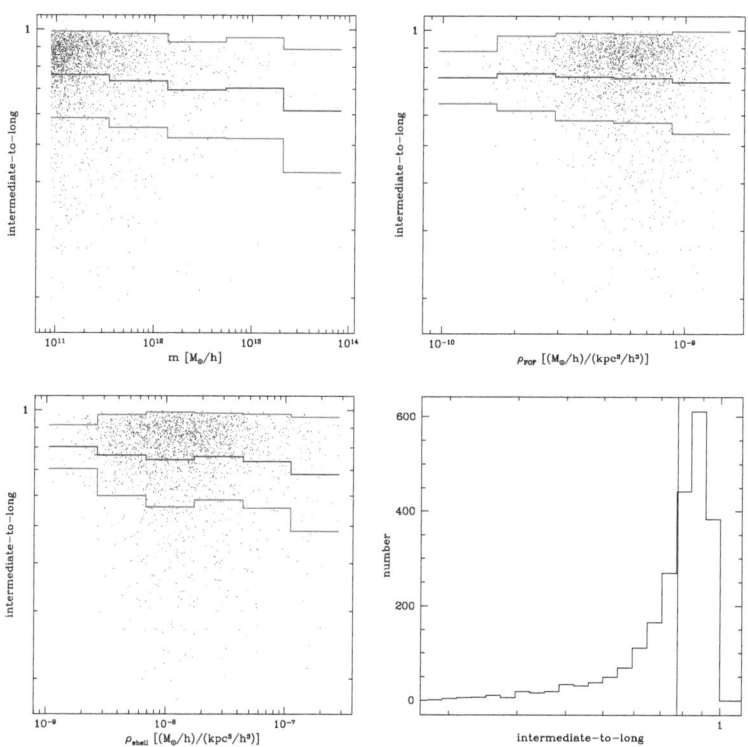

Figure 2.13: As figure 2.12 but for the intermediate-to-long axis ratio. We find no correlation of this ratio with the environment, but a weak anticorrelation with the halo mass.

2. Galaxies and Environment

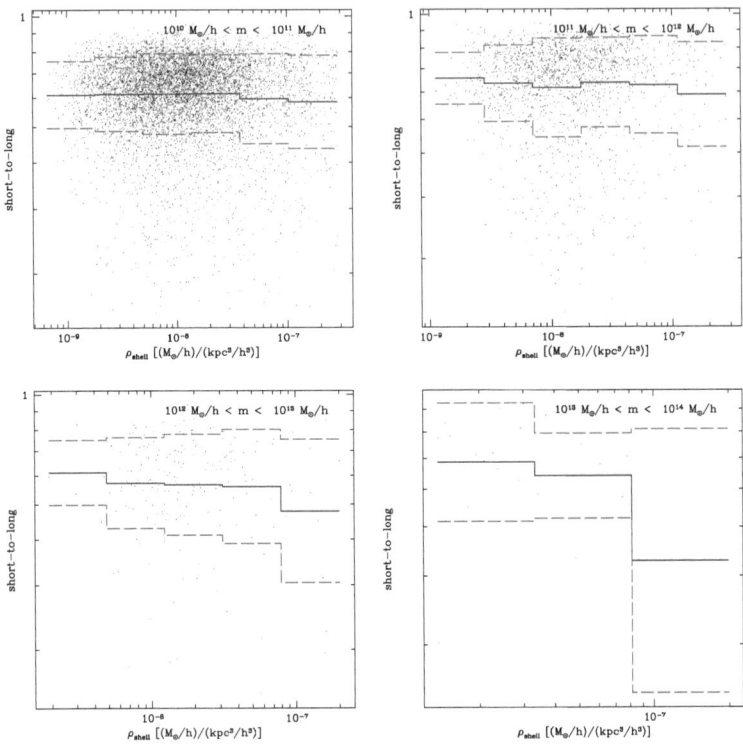

Figure 2.14: The correlation of the short-to-long halo axis ratio in the mass bins 10^{10} M$_\odot$/h to 10^{11} M$_\odot$/h (upper left), 10^{11} M$_\odot$/h to 10^{12} M$_\odot$/h (upper right), 10^{12} M$_\odot$/h to 10^{13} M$_\odot$/h (lower left) and 10^{13} M$_\odot$/h to 10^{14} M$_\odot$/h (lower right). We find no correlation and thus that the correlation seen in Figure 2.12 is a result of the correlation between mass and density.

2.4 Dark Matter Properties

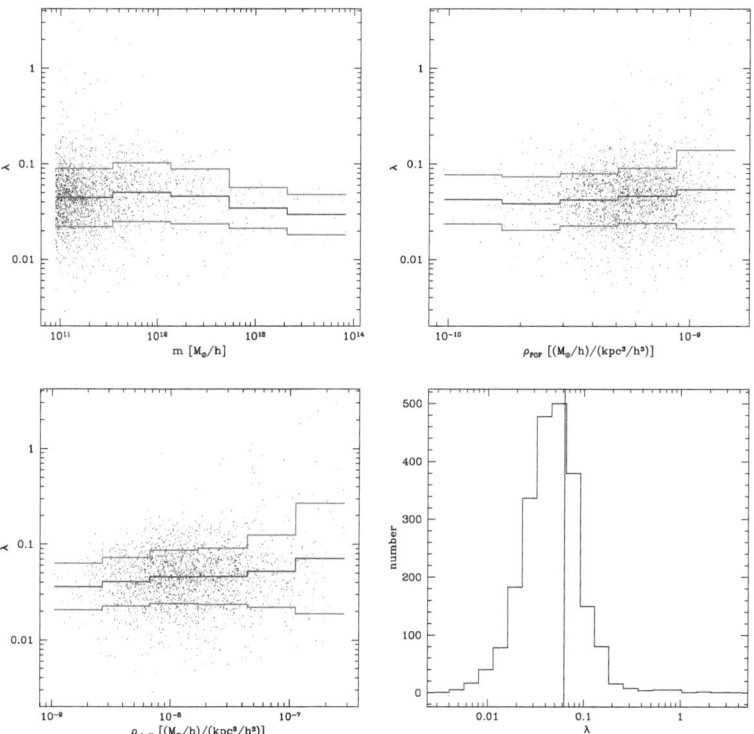

Figure 2.15: Correlation of the M3 spin parameter λ with the halo mass mass (upper left) and with the density (upper right : "FOF"-method, lower left, lower left: shell method). The distribution of the λ-values as well as the mean value (0.062) are shown in the lower right panel.

2. Galaxies and Environment

mass bins using the density evaluation in shells as shown in Figure 2.17. The correlation of the formation redshift with the environment from Figure 2.16 is thus just a result of the established correlation between halo mass and formation redshift. Our finding is thus consistent with that of Lemson & Kauffmann (1999).

2.4.6 Last major merger

As a last DM property we examine the mass and density dependence of the last major merging event, i.e. the last redshift z_{merger} where the second most massive progenitor of a halo was more than 0.3 times as massive as the most massive one. In the theory of galaxy formation this measure is important because it marks the transition region between a major merger where the cold gas is consumed in a large star-burst to produce a galaxy bulge and the regime where the smaller galaxy is just accreted onto the larger system without destroying its structure.

We again use all haloes with more than 500 DM particles for the analysis. We find no correlation between the last major merging event and the mass or the local environment of a halo (Figure 2.18). This is consistent with the analytical result of Lacey & Cole (1993).

2.5 Galaxies

The very high mass resolution of M3 together with the fact that the HR region of the simulation series had been computed within the full cosmological context makes these simulations predestined for the application of a semi-analytical model of galaxy formation (section 1.3).

The mass resolution allows the resolution of the DM haloes of dwarf galaxies. In order to faithfully compute the properties of larger galaxies, i.e. galaxies of the size of the Milky Way, the correct evolution of the progenitors is essential.

Due to the high quality of the resulting galaxy catalogue, several projects using the data of the simulation have been carried out (Ciardi, Stoehr & White 2003b; Ciardi, Ferrara & White 2003a) or are in preparation.

We extend our environmental analysis of the DM haloes above to the model galaxies produced by the semi-analytic model. In this section we first describe the semi-analytic model of galaxy formation that we have used. We then show the global properties of the produced galaxy population, i.e. the galaxy luminosity function, the Tully-Fisher-Relation and the global star-formation history. We finally correlate individual galaxy properties with the local environment in sections 2.5.6.1 and 2.5.6.2.

2.5 Galaxies

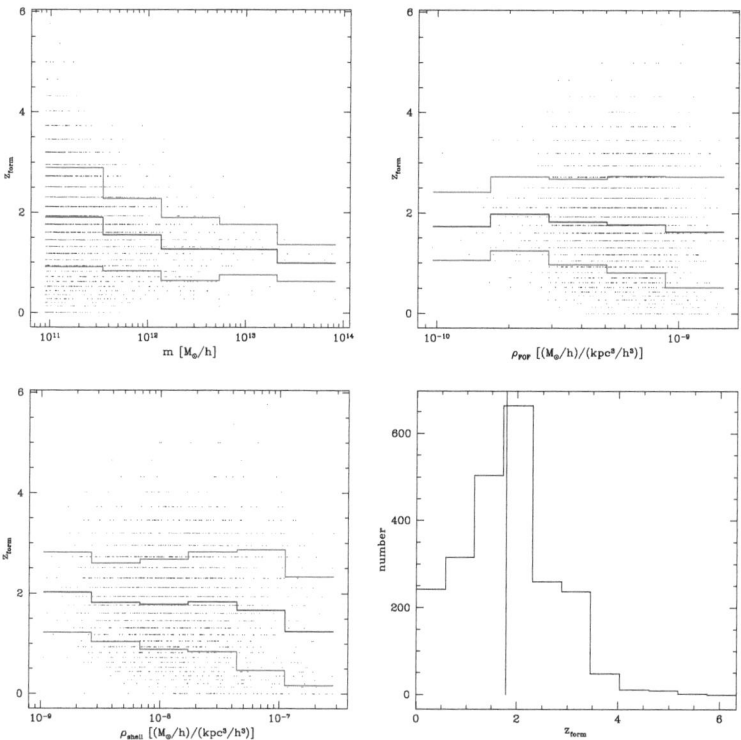

Figure 2.16: As the previous plots, this time showing the halo formation times. The formation time is strongly correlated with mass. This correlation is the reason for the trends observed in the upper right and lower left panel as can be seen from Figure 2.17.

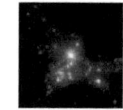

2. Galaxies and Environment

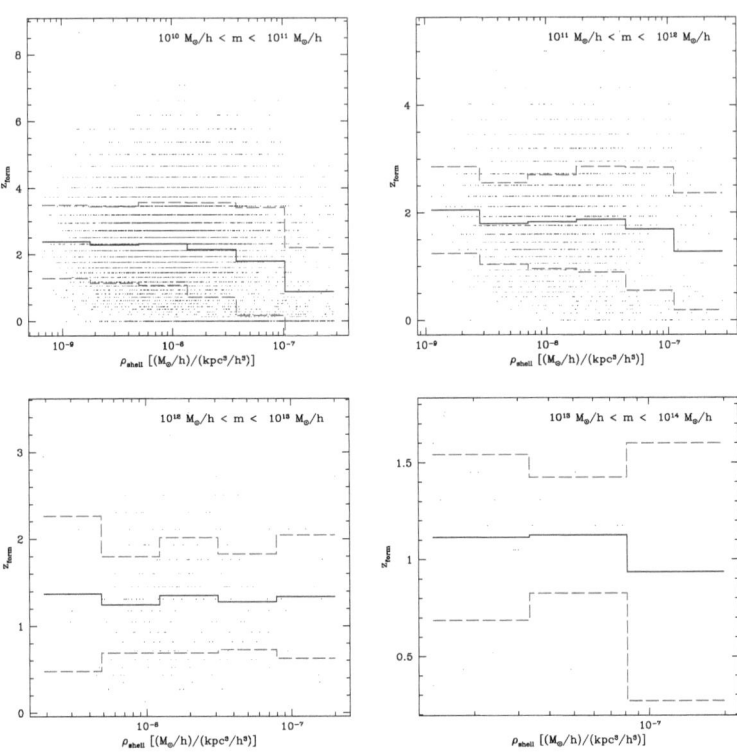

Figure 2.17: Correlations of the formation time with the density (evaluated with the shell-method). The figures correspond to the mass ranges 10^{10} M$_\odot$/h to 10^{11} M$_\odot$/h (upper left), 10^{11} M$_\odot$/h to 10^{12} M$_\odot$/h (upper right), 10^{12} M$_\odot$/h to 10^{13} M$_\odot$/h (lower left) and 10^{13} M$_\odot$/h to 10^{14} M$_\odot$/h (lower right).

2.5 Galaxies

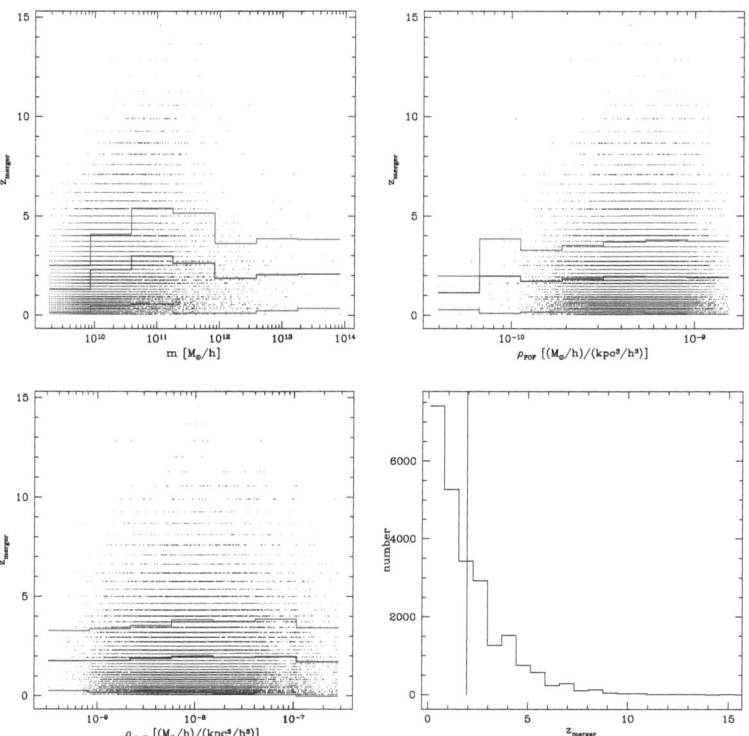

Figure 2.18: Correlation between mass and the last major merging event (mass-ratio of the progenitor haloes larger than 0.3). There is a no correlation with the halo mass or with the environment.

2.5.1 Galaxy formation model

We use the version of the semi-analytic modelling that includes the treatment of substructure haloes developed by Volker Springel (Springel et al. 2001a) following the model of Kauffmann et al. (1999).

For all DM haloes identified by the FOF group finder, gravitationally bound substructure haloes are identified with SUBFIND (Springel et al. 2001a). This selects candidate substructure haloes by lowering a density threshold starting from the particle with the highest local halo density. For this, the particles are sorted according to density. If during the lowering of the threshold particles are treated that are close to the already processed ones, they are attached to corresponding substructure candidate. If the particle under consideration however is separated by any of the processed particles by more than a few softening lengths, then this particle is taken to be the seed of a new substructure halo. Once all particles are processed and all substructure candidates are determined, substructure particles that are not gravitationally bound are removed from the subhaloes. This is repeated iteratively until the gravitationally bound substructure is determined. (Springel et al. 2001a) show that a semi-analytic galaxy formation model that correctly follows the evolution of the substructure haloes performs significantly better than one which does not follow substructure.

After the substructure haloes are detected, a merging tree is built by following the simulation particles from one output to the other. A halo in output $i-1$ is defined to be a *progenitor* of a halo in output i if the second contains more than half of the particles of the first. A subhalo in output $i-1$ is progenitor of a subhalo in output i, if more than half of the $N_{link}=10$ gravitationally most bound particles end up in the new subhalo.

The cooling of the primordial gas is computed following White & Frenk (1991). A singular isothermal sphere density profile is assumed

$$\rho_g(r) = \frac{M_{hot}}{4\pi R_{vir} r^2}. \tag{2.6}$$

For this profile the virial temperature, virial mass and virial circular velocity are

$$T = 35.9\ (V_{vir}/\text{km s}^{-1})^2, \tag{2.7}$$

$$M_{vir} = \frac{100\ H^2 R_{vir}^3}{G}, \tag{2.8}$$

$$V_{vir}^2 = G\frac{M_{vir}}{R_{vir}}. \tag{2.9}$$

$$\tag{2.10}$$

The total gas mass associated with the DM mass of the halo is set as

$$M_{gas} = f_b M_{vir}, \tag{2.11}$$

where the baryon fraction f_b is a parameter of the model. With the dynamical time

$$t_{dyn} = \frac{R_{vir}}{V_{vir}}, \qquad (2.12)$$

the cooling time from Sutherland & Dopita (1993)

$$t_{cool}(r) = \frac{\frac{3}{2}kT}{\bar{\mu}\, m_p\, n_e(r)\, n_t(r) \Lambda(T,Z)}, \qquad (2.13)$$

can be converted into the cooling rate

$$\frac{dM_{cool}}{dt} = \frac{M_{hot}}{R_{vir}} \frac{r_{cool}}{2\, t_{dyn}}, \qquad (2.14)$$

which gives the rate at which the shock heated gas in the halo cools to temperatures at which the gas clouds collapse and can turn on star formation (see the definition of r_{cool} below).

The maximum rate at which gas can settle on the central galaxy however can be approximated by

$$\frac{dM_{accr}}{dt} = \frac{M_{hot}}{2\, t_{tdyn}} \qquad (2.15)$$

which is the limiting rate if the cooling time is so short that no dynamical equilibrium is reached. The minimum of the two rates is used to determine the amount of gas that is available for star formation.

2.5.1.1 Cooling

The cooling radius r_{cool} is defined as the distance from the halo centre at which the cooling time t_{cool} equals the dynamical time t_{dyn}. For gas that consists of H^+, He^{2+}

2. Galaxies and Environment

and e^- follows:

$$Y = \frac{4\,n_{He}}{n_H + 4\,n_{He}} \tag{2.16}$$

$$n_e = n_H + 2\,n_{He} = \frac{(2-Y)}{2(1-Y)} \tag{2.17}$$

$$n_t = n_H + n_{He} = \frac{4-3Y}{4(1-Y)} \tag{2.18}$$

$$n_{He} = \frac{Y}{4(1-Y)} n_H \tag{2.19}$$

$$n = n_{He} + n_e + n_t = \frac{8-5Y}{4(1-Y)} n_H \tag{2.20}$$

$$\bar{\mu} = \frac{4}{8-5Y} \tag{2.21}$$

$$\rho_g = \bar{\mu}\,m_p\,n \tag{2.22}$$

$$t_{cool} = \frac{K\,T\,\bar{\mu}\,m_p}{\Lambda_{norm}(T,Z)\,\rho_g(r)} \frac{3\,(8-5Y)}{(2-Y)(4-3Y)} \tag{2.23}$$

$$t_{dyn} = t_{cool} \tag{2.24}$$

$$r_{cool}^2 = \frac{R_{vir}}{V_{vir}} \frac{M_{hot}\,\Lambda_{norm}(T,Z)}{K\,T\,m_p\,4\pi\,R_{vir}} \frac{(2-Y)(4-3Y)}{3\,(8-5Y)} \tag{2.25}$$

With a He-fraction of $Y = 0.25$ this gives:

$$r_{cool}^2 = 0.28086\,\frac{R_{vir}}{V_{vir}} \frac{M_{hot}\,\Lambda_{norm}(T,Z)}{K\,T\,m_p\,4\pi\,R_{vir}} \tag{2.26}$$

Cooling is suppressed in substructure haloes and in haloes with circular velocities larger than 350 km/s accounting for the fact that cooling flow clusters do not show star formation at the rate computed above (Kauffmann et al. 1999).

2.5.1.2 Star formation

The cold gas turns into stars at a rate proportional to the mass of the cold gas available and inversely proportional to the dynamical time of the halo:

$$\frac{dM_*}{dt} = \alpha\,\frac{M_{cold}}{t_{dyn,*}} \tag{2.27}$$

with α being the star formation efficiency. The dynamical time of the stellar component $t_{dyn,*}$ is fixed at

$$t_{dyn,*} = \frac{0.1\,R_{vir}}{V_{vir}}. \tag{2.28}$$

this corresponds to assuming an extension of the stellar component of $0.1\,R_{vir}$.

2.5.1.3 Feedback

A fraction of the newly formed stars explode as supernovae and energy is released to the surrounding medium. Many feedback schemes have been studied (see Benson et al. (2003) for a resume). We use the original implementation from Kauffmann et al. (1999).

$$\frac{dM_{reheated}}{dt} = \frac{4}{3} \epsilon \frac{\eta_{SN} E_{SN}}{V_{vir}^2} \frac{dM_*}{dt} \qquad (2.29)$$

In the "ejection" model we use, the gas is ejected from the halo and is re-incorporated into the halo only on the dynamical time scale (see Springel et al. (2001a) for a detailed description).

2.5.1.4 Merging

Starting from the earliest output – at $z = 20$ for the M-series – a *central* galaxy is assigned to each newly identified halo. The amount of gas cooling, star formation and feedback that happens in the timespan between this and the next output is computed in 50 substeps that are equally spaced in time. The galaxy is attached to the most bound particle of the FOF halo. Each galaxy has two distinct "containers" for stars and star formation: the bulge component and the disk component. The cold gas and the stars that form from the cold gas in a halo are attributed to the disk component of the central galaxy.

The most massive galaxy attached to a FOF particle of the halo or subhalo is flagged as its *central galaxy* and is newly attached to the most bound particle of that group or subgroup at each output. Less massive galaxies within the halo or subhalo, i.e. *satellite galaxies*, will merge with the central galaxy on the dynamical friction time-scale

$$t_{friction} = \frac{1}{2} \frac{f(\epsilon)}{C} \frac{V_c\, r_c^2}{G\, M_{sat}\, \ln(\Lambda)} \qquad (2.30)$$

where f_e is set to 0.5 (Tormen 1997), $\Lambda = 1 + M_{vir}/M_{sat}$ and $C = 0.43$. The radius r_c is set to the virial radius of the halo when the satellite first fell onto it.

If the mass ratio of the stellar masses of the merging galaxies, is smaller than a threshold of 0.3 then the merging event is called a *minor merger*. In this case the stellar components of the satellite's disk and bulge are added to the disk and bulge of the central galaxy, respectively.

If the ratio of the stellar masses is larger than 0.3, the galaxies are supposed to undergo a star burst where all of the cold gas is transformed into stars. In addition, all stars of the two systems are transferred to the bulge of the new central galaxy which thus becomes an elliptical galaxy. Through further gas cooling it can grow a disk again later.

2. Galaxies and Environment

The photometric properties of the stars in the bulge and disk component are computed using the stellar population synthesis model developed by Bruzual A. & Charlot (1993).

A more detailed discussion of the semi-analytic model can be found in Kauffmann, White & Guiderdoni (1993); Kauffmann, Colberg, Diaferio & White (1999) and Springel et al. (2001a).

2.5.1.5 Model parameters

Semi-analytic models contain by construction a number of free parameters. The values of the free explicit parameters of the semi-analytic model presented above are shown in Table 2.4. These values were used throughout this work. They differ from the ones used by Springel et al. (2001a) only in the value of the baryon fraction. Instead of a value of 15%, a value of 12% provides for our simulations better fits to the observations.

Major-merger threshold	0.3
Baryon fraction f_b	0.12
Supernova-number per mass $[(10^{10} \times M_\odot/h)^{-1}]$ η_{SN}	0.005
Reheating efficiency ϵ	0.1
Star-formation efficiency α	0.05
Metal-yield	0.02

Table 2.4: Free parameters of the semi-analytic model. The Metal-yield is the fraction of mass of the stars that is made of elements heavier than He. The metalicity of the gas serves as input for the computation of the metal-dependent cooling function.

The number of actual free parameters is very small, especially as for all but the two efficiencies observations or simulations narrow the possible ranges. Note however that there is significant amount of freedom in the modelling of the different processes. The "Durham" semi-analytic model of Cole et al. (2000) for example assumes a different dynamical time for the star-formation.

In addition, other physical processes, like photoionisation, other forms of supernova feedback or a colour correction model accounting for the effects of dust can be added to overcome weaknesses of the model. The danger of this approach however is to introduce many new parameters and recipes that might better fit observations but do not necessarily provide deeper physical insight.

2.5 Galaxies

For the purpose of this chapter, the evaluation of the environmental dependence of the galaxy properties, the semi-analytic model described above reproduces the observations accurately enough.

2.5.2 Luminosity function

Although reproducing the luminosity function, i.e. the number density of galaxies as a function of their luminosity seems an easy task, so far all semi-analytic models predict both a faint-end slope that is too steep and well as too many bright galaxies. Figure 2.19 shows the luminosity functions of M0, M1, M2 and M3 together with the B_J-band luminosity function from the Sloan Digital Sky Survey (Blanton et al. 2001) (dashed) as well as the luminosity function from the 2dFGRS survey (Folkes et al. 1999) (dotted). The B_J band is (Metcalfe et al. 1995)

$$B_J = B - 0.35\,(B - V) \tag{2.31}$$

The survey data were kindly provided by Hughes Mathis.

The luminosity functions converge very well. The bright end and the "knee" of the luminosity function reasonably match the observed relation. The abundance of faint galaxies however is too large by about a factor of 5 at the limit of the resolution of M3. A careful discussion on the problem of the faint-end slope of semi-analytic galaxy formation models can be found in Benson et al. (2003). These authors conclude that supernova feedback and thermal conduction are - despite substantial progress - still poorly understood.

Our results of the luminosity function are similar to those in other work (Kauffmann et al. 1999; Springel et al. 2001a; Mathis et al. 2002). Note however, that modelling photoionisation (Somerville et al. 2001; Benson et al. 2003) can bring simulations in much closer agreement with the observed faint end of the luminosity function.

2.5.3 Tully-Fisher-relation

The Tully-Fisher relation is the correlation between a galaxy's rotation velocity, and its intrinsic luminosity. This relation is a prominent method of determining distance of galaxies.

Figure 2.20 shows the I-Band Tully fisher relation as observed by Giovanelli et al. (1997)

$$M_I - 5\,\log(h) = -21.00 - 7.68\,(\log(W) - 2.5) \tag{2.32}$$

(solid line) for the four simulations. Here M_I is the I-band magnitude of the galaxy and W the velocity dispersion.

2. Galaxies and Environment

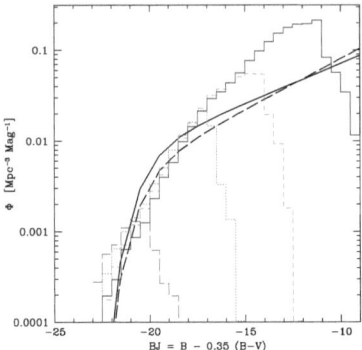

Figure 2.19: Luminosity function of the galaxies in M0 (long dashed), M1 (dot-dashed), M2 (short-dashed) and M3 (solid). The thick lines show the luminosity functions as observed by the Sloan Digital Sky Survey (Goto et al. 2002) (solid) and the 2dF Survey (Folkes et al. 1999) (dashed).

The dots show the values for the simulated central galaxies. The stellar disks have been assumed to rotate with the circular velocity v_{200} of the DM halo, hence, $w = 2v_{200}$. Whereas the slope of the relation is very well reproduced, the zero point is significantly offset.

Although the Tully-Fisher relation seems a good means to compare observed data to simulations, the uncertainty in how the rotation velocity of the stellar disk relates to the circular velocity of the DM halo makes the comparison less than straightforward. Figure 2.21 shows the relation where 85% of v_{200} of the DM halo has been used. Our simulated Milky Way rotation curves in Figure 4.2 show that this value corresponds approximately to the rotation velocity of the DM halo at the position of the Sun at 8 kpc from the galactic centre.

2.5.4 Star formation rate

We extend the range of times studied from the measurements at $z = 0$ to earlier times: The star formation history of the Universe was first compiled by Madau et al. (1996). Figure 2.22 shows the star formation history of our four simulations together with the dust-corrected data taken from Somerville, Primack & Faber (2001). We find reasonable

2.5 Galaxies

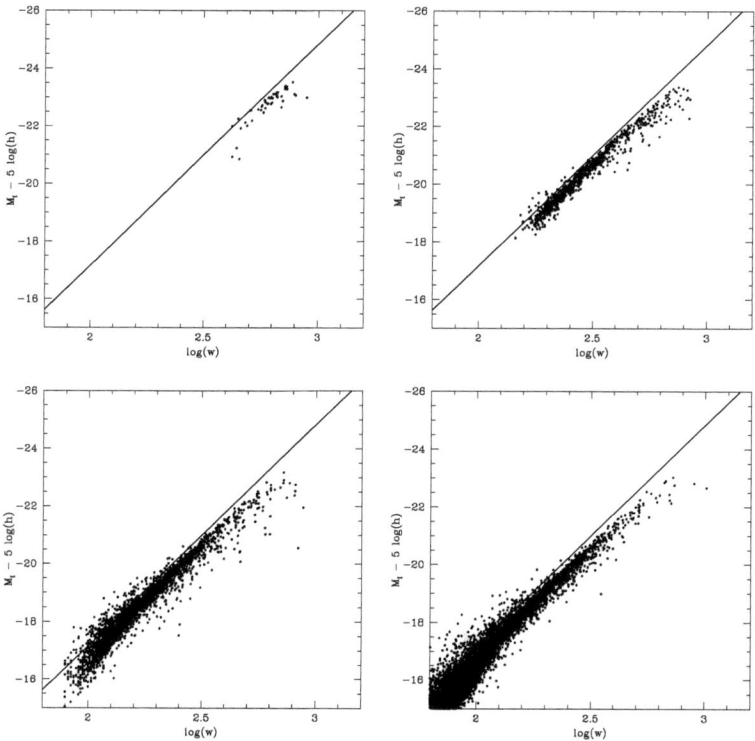

Figure 2.20: I-Band Tully-Fisher relation for the four simulations M0 (upper left), M1 (upper right), M2 (lower left) and M3 (lower right). Although the slope of the relation is rather well reproduced, there is a significant offset with respect to the observed relation (Giovanelli et al. 1997).

2. Galaxies and Environment

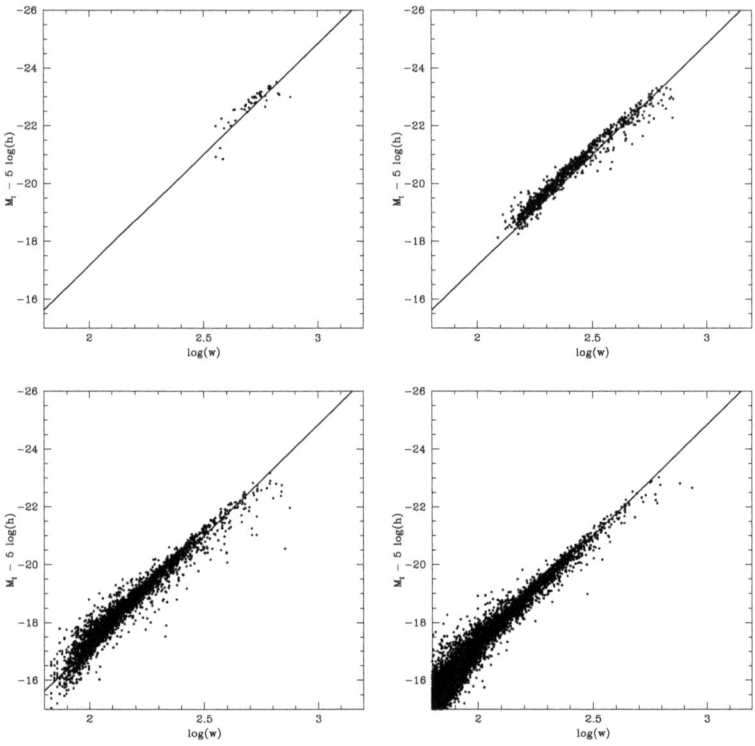

Figure 2.21: As figure 2.20 but this time using $v_{rot} = 0.85\ v_{200}$. This scaling corresponds roughly to the ratio of the circular velocity at the position of stellar disk versus v_{200} in the Milky Way.

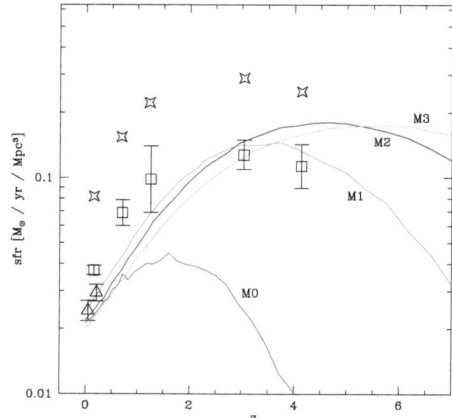

Figure 2.22: Global star-formation history of the four simulations. The data are taken from Somerville, Primack & Faber (2001) who applied dust and extinction corrections to the original data from Gronwall (1998); Tresse & Maddox (1998) (triangles) and Treyer et al. (1998); Cowie et al. (1999); Steidel et al. (1999) (boxes). Crosses show the "maximally dust corrected values" from Somerville, Primack & Faber (2001).

agreement with the data and the result from Somerville, Primack & Faber (2001). The curve is well below the maximally dust corrected data points (crosses).

From the curves, the approximate redshift of convergence can be determined for each of the simulations. Approximately, M0 has converged at only $z = 1$, M1 has converged at $z = 2.5$, and M2 has converged at $z = 5$. Extrapolating this series we expect the star formation history of M3 to have converged at about redshift $z = 11$ (see also (Ciardi, Stoehr & White 2003b)).

2.5.5 Galaxy distribution

Figures 2.23 to 2.26 show the galaxy distribution of M3 together with that of galaxies in the cluster simulation S3 performed by Springel et al. (2001a). In order to allow a direct comparison between both simulation series, the baryon fraction f_b adopted in Springel

2. Galaxies and Environment

et al. (2001a) has been used throughout.

The "galaxies" that are plotted over the dark matter distribution (blue) have areas proportional to their B-band luminosity. They are colour coded according to their "Hubble type" T. It was derived from the B-band bulge-to-disk luminosity ratio (Springel et al. 2001a)

$$m_{bulge} - m_{total} = 0.324 \, (T+5) - 0.054 \, (T+5)^2 + 0.0047 \, (T+5)^3 \qquad (2.33)$$

Elliptical galaxies are plotted in red ($T < -2.5$), spiral galaxies in yellow ($T > 0.92$) and S0s in green ($-2.5 < T < 0.92$).

In Figures 2.23 to 2.25 all galaxies of the catalogues are shown. Figure 2.26 shows only galaxies with stellar masses larger than 10^9 M$_\odot$/h. This mass threshold corresponds approximately to the stellar mass of the Small Magellanic Cloud.

The figures show that the descendants of the first stars are to be found in the centres of the large galaxy clusters today. They also show, that galaxy formation is strongly suppressed in the large underdense regions (see also Mathis & White (2002)).

2.5.6 Galaxies and environmental effects

In this section we finally analyse the correlation of two of the properties of our model galaxies with their local environment, i.e. the local overdensity of their host DM halo. We only use galaxies with stellar masses larger than 10^9 M$_\odot$/h. In addition, we require them to reside in in haloes with more than 100 simulation particles. These choices assure that the galaxies we plot are free from resolution effects.

2.5.6.1 Bulge-to-total luminosity ratio

The bulge-to-total luminosity ratio is a direct measure of the type of the galaxy as large values (close to 1) indicate bulge-dominated, i.e. elliptical, galaxies and small values (close to zero) disk-dominated, i.e. spiral galaxies. The ratio is rather easily observable for many galaxies.

Figure 2.27 shows the bulge-to-total luminosity ratio as a function of the stellar mass of the galaxy mass and local density evaluated at the position of the parent DM halo. We find no correlation with the stellar mass. This is a non-trivial result. There are many galaxies within a galaxy cluster that merge with the massive central galaxy. The probability of a major merger event should therefore be larger than for two satellite galaxies within this halo. Massive galaxies should have larger bulge-to-total ratios than smaller galaxies. On the other hand, in systems where the cooling cut-off is not reached, cooling onto the stellar disk only happens for the central galaxy. If no merging were present, the bulge-to-total luminosity of these galaxies would decrease.

2.5 Galaxies

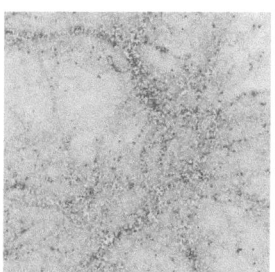

Figure 2.23: Distribution of galaxies in S3 (upper panel) and M3 (lower panel) at redshift $z = 5$. The thickness of the boxes is 10 Mpc/h, the side lengths are 10 Mpc/h and 20 Mpc/h for the S3 and the M3 box, respectively. Shown are all galaxies down to the resolution limit of M3 (10 particle DM haloes). The galaxy symbols (ellipticals red, S0's green and spirals yellow) have sizes proportional to their B-band luminosities. The DM distribution is shown in blue-scale.

2. Galaxies and Environment

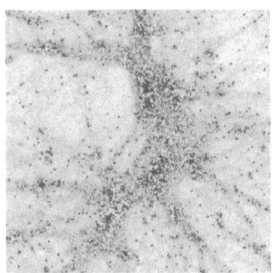

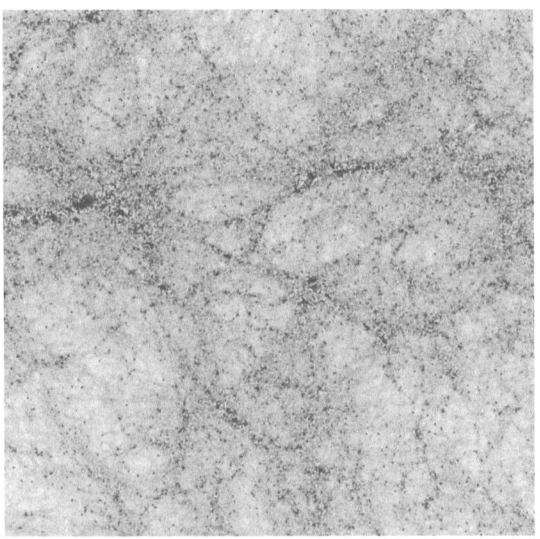

Figure 2.24: As Figure 2.23 at $z = 2$

2.5 Galaxies

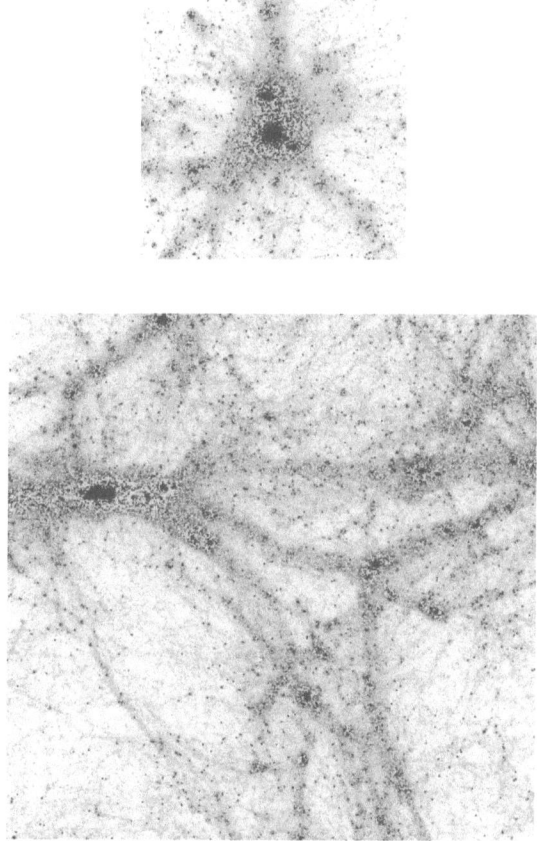

Figure 2.25: As Figure 2.23 at $z = 0$

2. Galaxies and Environment

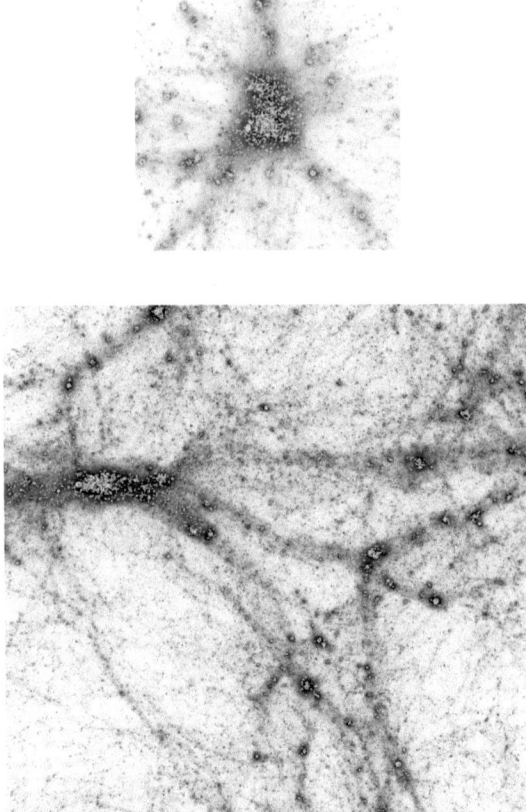

Figure 2.26: As Figure 2.25 but this time restricting to galaxies in DM haloes more massive than 10^9 $M_\odot/h$ corresponding approximately to the size of the small Magellanic cloud.

The upper right and the lower left panel of Figure 2.27 show the "morphological segregation". Vertical concentrations of galaxies arise because all galaxies in a halo get the same density assigned. We find that galaxies in high density environments tend to be more elliptical. This correlation is very strong. The morphological segregation can be observed directly in galaxy clusters (Dressler 1980) or by measuring the two point correlation function of galaxies of different types (Madgwick et al. 2003).

2.5.6.2 B-V colour index

A very easily accessible observable of a galaxy is its colour. The colour is quantified by the magnitude difference of two bands, for example the B and V (bands centred on 440 nm and 540 nm, respectively). The B-V values correspond roughly to the colours

$$B - V = +1.41 \rightarrow Red$$
$$B - V = +0.82 \rightarrow Orange$$
$$B - V = +0.00 \rightarrow White$$
$$B - V = -0.29 \rightarrow Blue$$

(2.34)

The correlations of the B-V colour index of the galaxies in our catalogue with their stellar mass and the local density are shown in Figure 2.28. A large population of faint galaxies is very red. These galaxies are satellite galaxies that have fallen into larger systems. As our model prevents gas from cooling in this case, no star formation occurs and the galaxies turn red.

The colour mass distribution is similar to the colour magnitude relation found in observations (Bell et al. 2003). However, instead of a slight reddening of the galaxies with increasing mass, we find a slight negative correlation. Kauffmann & Charlot (1998) showed, that when modelling metalicity dependent stellar evolution, the correct slope of the colour magnitude relation can be obtained.

We find a strong trend of galaxies in high-density environments to be redder. Our simulation thus reproduces the correlation found in observations (Kodama et al. 2001). In addition, our model predicts a bimodal colour distribution which is, too, observed (Bell et al. 2003).

2.6 Conclusions

We have performed a series of high-resolution simulations of a large region of the Universe within the full cosmological context. State-of-the-art techniques were used to resimulate

2. Galaxies and Environment

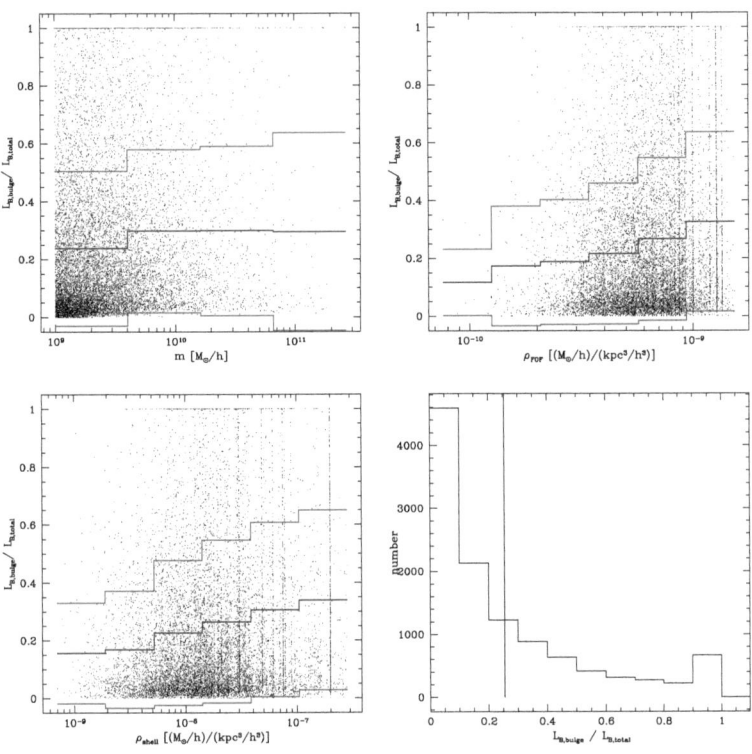

Figure 2.27: Bulge-to-total luminosity for galaxies in M3 with stellar masses of more than 10^9 M$_\odot$/h.

2.6 Conclusions

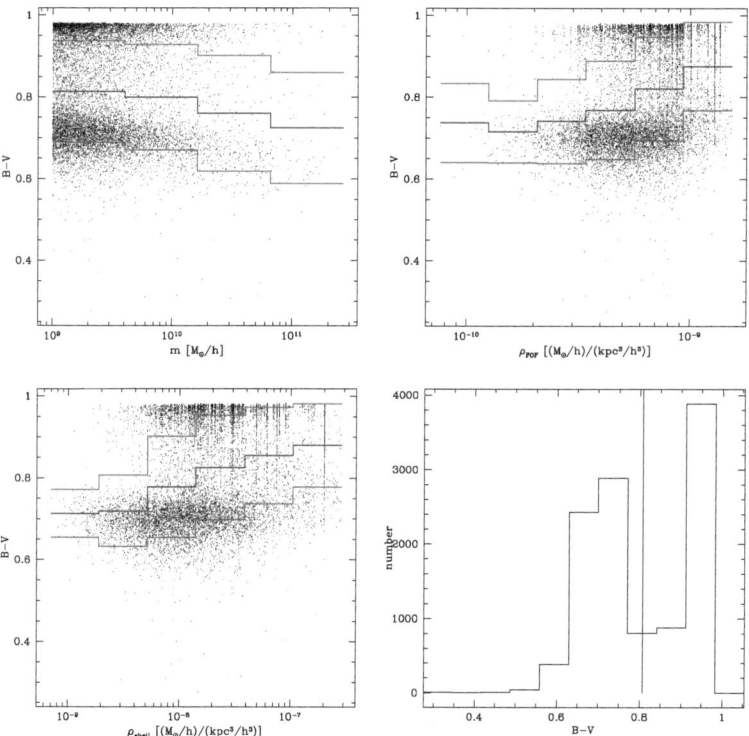

Figure 2.28: B-V colour index of galaxies with stellar masses larger than 10^9 $M_\odot/h$. The model predicts a bimodal colour distribution and the correct trend of the colour-density relation.

a region that was carefully selected to have mean properties. The simulated region has very high mass resolution, is a representative patch of the Universe and evolves under the influence of the large scale tidal field at the same time.

The simulation series is or has been used for studies of reionization (Benedetta Ciardi), metal enrichment in the intergalactic medium through supernova driven winds (Serena Bertone), analysis of substructure in DM haloes (Gabriella daLucia), testing of a new tessellation method (Willem Schaap), the distribution of satellites around large galaxies in the 2dF Survey (Peder Norberg), the mass accretion histories of haloes and their concentration parameters (Frank van den Bosch), the spin-spin correlation of DM haloes (Tom Theuns) and predictions for observations of the Lyman-α forest (Vincent Desjaques and Adi Nusser).

We used the simulation as parent simulation for the ultra-high resolution simulations analysed in chapter 3 and chapter 4.

In this chapter we have studied the influence of the matter overdensity of the local environment of DM haloes and model galaxies on their properties. We have proposed a new method for the estimation of this overdensity, that completely discards mass that is bound in collapsed objects. Using two different methods for the estimation of the local overdensity of the matter distribution allowed us to verify the robustness of the correlations between the overdensities and the properties of the studied objects.

We first analysed the properties of collapsed DM haloes. We found that the concentration parameter of the NFW profiles fitted to our DM haloes is correlated with the matter density in the local environment of the haloes. Haloes in high density environments tend to have larger concentration parameters. In addition, we have shown that the spin parameter of the haloes exhibits a density dependence. Haloes in denser environments tend to be more rotationally supported than their low density environment counterparts. This result is consistent with the analyses of Heavens & Peacock (1988) and Steinmetz & Bartelmann (1995) and the findings of Lemson & Kauffmann (1999).

We have found no correlations of the halo axis-ratios, formation times or times of the last major merging events with the local overdensities and thus confirmed the results of Lemson & Kauffmann (1999). The fact that the formation histories of the haloes do not depend on the local environment is a prediction of Bond et al. (1991). We compared the mass functions of the DM haloes with the analytical prediction of Sheth et al. (2001b) extending its previously tested range by two orders of magnitude. We found that the prediction matches well our results at redshift $z = 0$. At higher redshifts, the mass function of the simulated DM haloes is significantly steeper.

We applied a semi-analytic model of galaxy formation to our simulation series and adjusted the free parameters to match the luminosity function of observed galaxies. Our model fails to reproduce the faint end slope of the luminosity function. Additional

physics (e.g. photoheating) has to be included into the model. We find that the global star formation history of the model galaxies for our best fitting parameters is in very good agreement with observations.

Finally we studied correlations of the bulge-to-total luminosity ratio and of the B-V colour index of the model galaxies with the galaxy-environment. We find a strong luminosity ratio and B-V dependence on environment, consistent with observations. The model also correctly predicts the trend of the colour-density relation found in observations. In addition, the bimodal distribution of galaxy colours is obtained.

2. Galaxies and Environment

3 Satellite Galaxies of the Milky Way

3. Satellite Galaxies of the Milky Way

3.1 Introduction

Blumenthal et al. (1984) and Davis et al. (1985) showed that properties of the Large Scale Structure of the Universe can nicely be explained when assuming that the dark matter (DM) consists of heavy non-relativistic particles. Since then, the model became more and more popular. Today the variant with a non vanishing cosmological constant Λ, the ΛCDM model, is the "standard" model of structure formation (see section 1.2.1). In this scenario the smallest structures collapse first and during the evolution of the Universe larger structures form by merging of the smaller Dark Matter haloes that have collapsed earlier.

For a DM halo of the size of the Milky Way, assuming simple recipes for gas cooling, star-formation, and stellar feedback (Kauffmann & White (1992), Kauffmann, White & Guiderdoni (1993), and also sections 1.3 and 2.5), the number of visible satellites predicted to orbit in the DM halo exceeds by far the actual number (Kauffmann, White & Guiderdoni 1993). Eleven satellites are known within the virial radius of the Milky Way. The main reason for this is the large number of smaller dark matter objects that are accreted during a Hubble time. This overabundance is directly related to the shape of the dark halo mass function (Sheth et al. (2001b), see also 2.4.1).

The same discrepancy appears at the faint end of the luminosity function of galaxies which is observed to be shallower than the prediction of simple models of hierarchical galaxy formation (White & Rees 1978; White & Frenk 1991). Kauffmann, White & Guiderdoni (1993) argued, therefore, that star formation in these small objects must be strongly suppressed, perhaps by the radiation processes discovered by Efstathiou (1992). As a result only the most massive Milky Way satellites would be visible, reconciling the model with the observations.

This suppression could be achieved a consequence of photoionisation in the early stages of the Universe (Bullock et al. 2000; Benson et al. 2002). Quasars and early galaxies emit strongly at energies high enough to reionise hydrogen, which had been neutral since the epoch of recombination some 300,000 years after the Big Bang (1.2.1). This ionising radiation not only reheats part of the gas of the intergalactic medium which could otherwise cool and form stars, but also reduces the cooling rate of this gas because less gas is in a molecular state in which cooling is efficient (section 1.3.1, Doroshkevich et al. (1967); Couchman & Rees (1986)). Reionisation is currently a very lively field of research (Wyithe & Loeb (2003); Baumann et al. (2003); Kaplinghat et al. (2003); Lamb & Haiman (2003); Doroshkevich et al. (2003) to cite a few).

As numerical simulations have been successfully used to study structure formation in the Universe, it seems natural to extend the technique to the study of single objects like the DM halo of the Milky Way. Such simulations can not only check the predictions

3.1 Introduction

of the simplified galaxy formation models in sofar as the survival of infalling satellites is concerned, but can also provide the orbits, sizes and internal structures of accreted subhaloes.

Although seemingly straight-forward, simulations of the Milky Way halo turned out to be very challenging (van Kampen 1995; Summers, Davis & Evrard 1995; Moore, Katz & Lake 1996). The simulated haloes were very smooth and infalling substructure was nearly completely destroyed, an effect that was called "over-merging problem". The solution to this problem finally came when the resolution of the simulations was dramatically increased (Tormen et al. 1998; Ghigna et al. 1998; Moore et al. 1998; Klypin et al. 1999; Ghigna et al. 2000). These high resolution simulations revealed a wealth of substructure within the dark matter haloes, accounting for roughly 5% to 10% of the mass of the main halo. This brought the picture sketched by the simple models of hierarchical galaxy formation into agreement with the numerical simulations. Still, the suppression of visible objects, through photoionisation for example, is necessary to reconcile the new simulations with observations.

The situation changed dramatically when Moore et al. (1998) and Klypin et al. (1999) published results on the internal structure of simulated subhaloes and claimed them to strongly disagree with observation. According to their estimates, the abundance of massive subhaloes was much greater than inferred from the Magellanic clouds and the nine Milky Way dwarf spheroidals.

To arrive at this conclusion, these authors compared the circular velocities of the subhaloes in their simulations with the central velocity dispersions of the stars at the centre of the satellite galaxies. Whereas the former are in the range of 20 to 60 km/s for the 10 to 20 most massive haloes, the latter are observed to be below 10 km/s in the smaller satellites. Simulated Milky Way haloes have typically more than a hundred subhaloes with circular velocities larger than 10 km/s.

Whereas the problem of the overabundance of visible subhaloes can be solved by the introduction of photo-heating, the internal structure of the simulated seemed to robustly disagree with the observations. The so called "substructure problem" or "substructure crisis" became one of the two major challenges to the ΛCDM model.

The second challenge, still remaining today, is the problem that the simulated DM haloes, at least the haloes that are not subsystems of larger haloes, have density profiles with steep inner slopes. Such profiles are termed "cuspy". Navarro, Frenk & White (1997) performed high resolution simulations and found these profiles to have "universal" form (see also Tormen, Bouchet & White (1997)). This form seems to be inconsistent with the near solid-body circular velocity curves observed in dark matter dominated dwarf galaxies (Flores & Primack 1994; Moore 1994; Burkert 1995; McGaugh & de Blok 1998; Firmani et al. 2001), although the extent of this problem is controversial, e.g. van

den Bosch & Swaters (2001). Some of the most recent studies (de Blok, McGaugh & Rubin 2001a) continue to claim that simulations can not be reconciled with observations.

In order to solve the substructure problem, several solutions have been proposed. Bode, Ostriker & Turok (2001a) suggested that the dark matter particles might not be "cold" but "warm" with masses around 1 keV. The idea is that due to their larger thermal velocities, these particles would have been able to wash out the density fluctuations at early times that lead to subgalactic structures today. This scenario seems to be ruled out by the recent results of the WMAP CMB experiment (Spergel et al. 2003; Yoshida et al. 2003).

A second suggestion was to allow collisional self-interaction of DM particles (Spergel & Steinhardt 2000; Yoshida et al. 2000). However, as Yoshida et al. (2000) show, self-interacting dark matter can not solve the substructure problem and their lensing studies favour *collisionless* dark matter.

In our study in this chapter we revisit the substructure problem. White (2000) emphasised that if the NFW profile is appropriate for subhalo structures, the maximum circular velocities of the subhaloes that host Milky Way dwarf galaxies can be much larger than the observed velocity dispersions. In order to quantify this effect, we perform a series of high resolution simulations of a relaxed Milky Way type halo in its full cosmological context. Instead of converting the central velocity dispersions of observed galaxies into halo circular velocities *ad hoc*, as was done by Moore et al. (1998) and Klypin et al. (1999), we use the observed stellar distributions to predict central velocity dispersions for the satellites sitting the potential wells of our simulated substructures. We then compare these values to the observed ones. Finally, we compare the velocity dispersion profiles we predict with those observed for Fornax and Draco, the only two Milky Way satellites where this information is available.

The remainder of this chapter is organised as follows: In section 3.2 we describe the process of the generation of the initial conditions of the simulation series and give the results for the main halo as well as for the subhaloes. The results include information necessary for the next chapter. In section 3.3 we summarise the observations that we use. Section 3.4 deals with the theory of statistical motion in a given potential. We compare the predicted velocity dispersions with the observed values in section 3.5 and conclude in section 3.6.

3.2 The simulated Milky Way

We have performed a series of dark matter simulations of a Milky Way sized halo (mass $\approx 2.1 \times 10^{12} M_\odot/h$) in its full cosmological context. The simulations are carried out in a ΛCDM cosmology. In the simulation with the highest mass resolution there are about

3.2 The simulated Milky Way

10^6 particles within the virial radius and the gravitational softening is 0.18 kpc/h. These simulations were then scaled down in length and in mass in order to match the peak of the circular velocity curve of the Milky Way (White & Springel 2000; Helmi & White 1999).

3.2.1 Initial conditions

As for the M-series simulations (section 2.2.1.3) we use the resimulation technique to achieve high mass resolution while preserving at the same time the tidal effects - and therefore the cosmological evolution - of a very large cosmological volume. After selecting reasonable candidates out of M3 (section 3.2.1.1) we extended the resimulation technique (section 3.2.1.2) in order to achieve the mass and length resolution mentioned above. In section 3.2.1.4 we discuss the contamination of low-resolution particles in the HR zone and describe the simulations performed in section 3.2.1.5.

3.2.1.1 Selection procedure

The Milky Way is a spiral galaxy with probably a rather quiet recent merging history. The last major merger is estimated to have happened probably around redshift $z = 2$ to 5 (Gilmore et al. (2002), and references therein). In addition, it is quite isolated and the nearest galaxy cluster is at a distance of about 20 Mpc (Virgo). The peak rotation velocity of the Milky Way has been measured to about 220 km/s. Our intention was to select a candidate a candidate for resimulation with properties as close as possible to the observed ones. It is possible to rescale the simulated objects in mass and radius (and so in characteristic velocity) by factors near unity provided the combination M/R^3 (which specifies the timescale) is kept fixed. The exact matching of the peak velocity was therefore the least important criterion.

In order to be able to reproduce a halo accurately in a resimulation, the object in the parent simulation has to have more than ≈ 500 particles (Stoehr 1999). Milky Way type haloes in the GIF ΛCDM, the best resolved full-box simulation with periodic boundaries however have only about 140 particles. We confirmed with tests that this number of particles was not sufficient using the techniques of Tormen, Bouchet & White (1997).

The best parent simulation available to us was therefore M3 (see section 2.2.1.3) with its mass resolution of 1.66×10^8 $M_\odot$/h where candidate Milky Way haloes contain roughly 10^4 particles. We required the candidate haloes to be positioned well inside the HR zone (i.e. closer than 15 Mpc/h from the centre), to have had their last major merger before $z = 1.2$, and not to be close to a galaxy cluster. From the four candidates identified (Table 3.1), candidate G-2 matched our requirements best.

3. Satellite Galaxies of the Milky Way

Candidate	G-1	G-2	G-3	G-4
Catalogue number (0 to n)	248	276	414	582
Redshift of last major merger	3.19	2.71	2.94	1.98
Distance to M3 centre [Mpc/h]	13	7	14	12
Circular velocity V_{200} [km/s]	219	205	182	151
Peak velocity V_{max} [km/s]	262	245	224	187
Virial radius R_{200} [kpc/h]	219	205	182	151
Mass M_{200} [$10^{12} \times M_\odot/h$]	2.45	2.0	1.42	0.8
Particle number N_{200}	14716	12010	8529	4796

Table 3.1: Properties of the four candidates in M3 that fulfilled our selection criteria. The candidate that matched our requirements best and that was used for the resimulation series described in this chapter was candidate G-2.

The closest large objects are a halo with 3 times and a halo with similar mass at distances of 3 Mpc and 4 Mpc, respectively. There is no major merger in the close future of the candidate at z=0. Figure 3.1 shows the density and velocity profiles. The agreement of the profiles with the analytical NFW fit for a concentration parameter $c = 11.2$ is remarkably good, taking account of the small particle number. Figure 3.2 shows the radial velocity profile and the mass-accretion history for this candidate. The merging history is shown in Figure 3.3.

The position of the candidate in the units of the simulation are

Resimulating an object in M3 which is itself a resimulation, required substantial changes to ZIC, the code used for the generation of the initial conditions (Tormen, Bouchet & White 1997).

3.2.1.2 Twolevel-ZIC

In addition to the steps for the generation of the initial conditions described previously (section 2.2.1 and Tormen, Bouchet & White (1997); Stoehr (1999)) an additional region

3.2 The simulated Milky Way

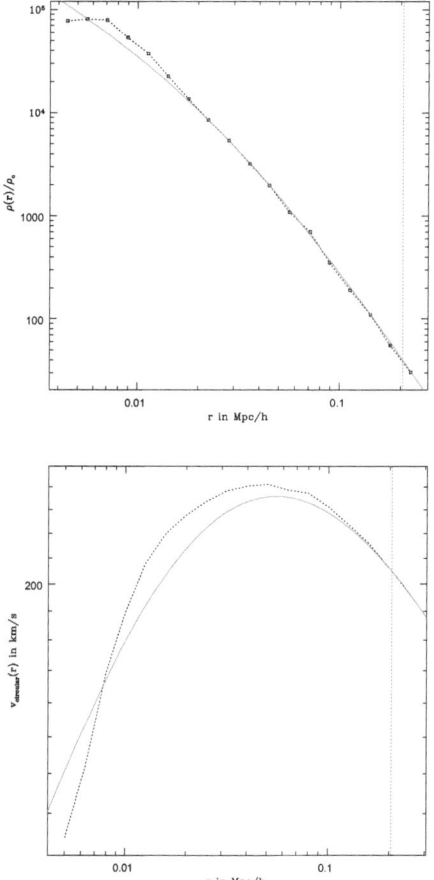

Figure 3.1: Upper panel: density profile of the candidate G-2 in units of the critical density. The solid line show a NFW profile with concentration parameter $c = 8.3$. The vertical line indicates the virial radius. Lower panel: Corresponding circular velocity profile. For a halo with this low particle number the agreement between the profile and the fit is remarkably good, indicating that the halo is relaxed. The gravitational softening is 1.4 kpc/h.

3. Satellite Galaxies of the Milky Way

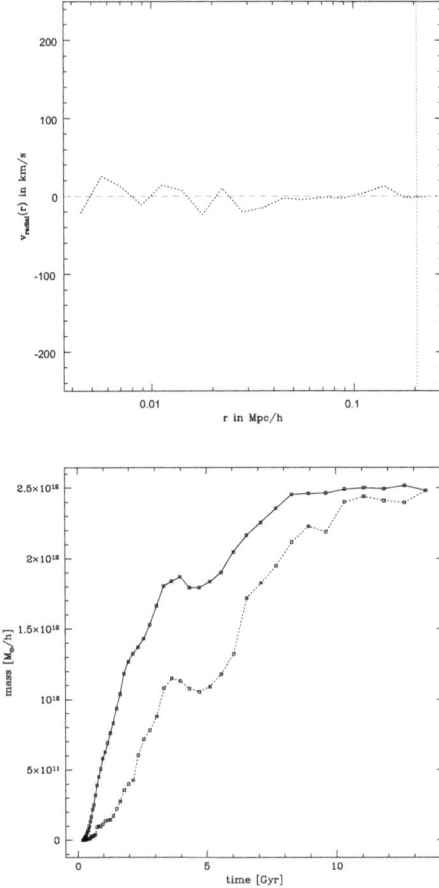

Figure 3.2: *Upper panel:* radial velocity within the virial radius. *Lower panel:* Mass growth history of the candidate halo G-2 in the M3 simulation. The solid line shows the mass of all progenitors of the final halo, the dotted line shows the mass of the most massive progenitor.

98

3.2 The simulated Milky Way

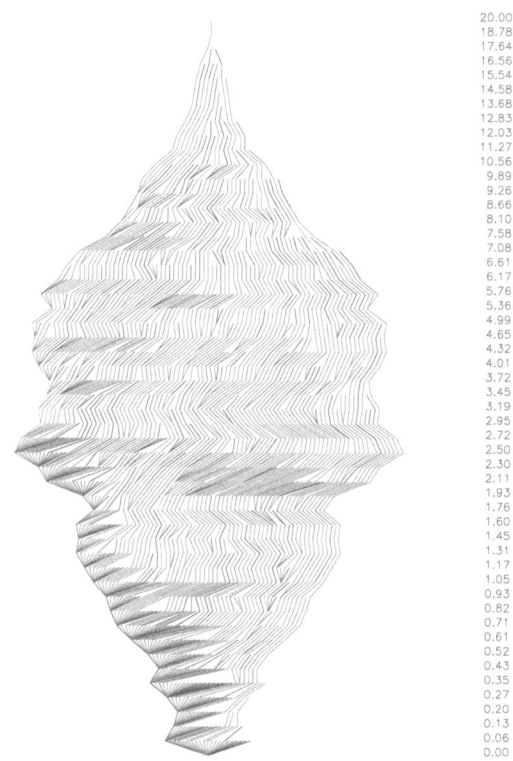

Figure 3.3: Merging history. The haloes at each output are sorted from left to right according to their particle number so that the leftmost line corresponds to the most massive progenitor. The redshifts of the outputs are shown on the right. For this plot, all haloes with more than ten particles were taken into account.

3. Satellite Galaxies of the Milky Way

	M3	GA0n	GA1n	GA2n	GA3n
x [Mpc/h]	5.9265	1.19995	1.38941	1.5901	0.4159
y [Mpc/h]	-11.8997	-4.78985	-4.74637	-4.72403	-5.0260
z [Mpc/h]	-2.2299	6.30782	6.49764	6.68449	6.0047

Table 3.2: Positions of the halo centres of the MW halo in the different simulations.

containing super-high resolution (SHR) particles had to be defined, set up and perturbed with a displacement field obtained by summing that of the LR particles, that of the HR particles, and a newly created field with spatial frequencies above the Nyquist frequency of the HR region. In addition, HR particles outside this new SHR region have to be regrouped on a spherical grid.

We added these features to the initial condition generation code ZIC and changed the code for tracking contaminating particles, in this case HR particles falling in the SHR zone, back to their original positions in M3.

For the high mass resolution we wanted to achieve (2×10^5 $M_\odot$/h) and the accuracy we required for structure on small scales, we decided to use the power spectrum generated with the publically available package CMBFAST (Seljak & Zaldarriaga 1996) instead of the analytical fit of Bond & Efstathiou (1987). In order to obtain exactly the structure of M3 however, the original power spectrum had to be used for the LR and HR particles. The CMBFAST power spectrum was only used to create the displacement field of the SHR particles. We checked, that the BE and the CMBFAST power spectrum match reasonably well at the Nyquist-frequency of the HR particles: the relative deviation of the transfer functions at this point is 31 %.

For the simulation series, the matching frequency of the perturbation modes for LR and HR particles was at 2.35 Mpc^{-1} and the matching of HR and SHR particles (corresponding to the Nyquist-frequency of the HR particles) at 17.5 Mpc^{-1}. The HR box and the SHR box were replicated 10 times in each direction. The Nyquist-frequency of the SHR particles was 163 Mpc^{-1} (Figure 2.3).

A problem in the interpolation routine of ZIC had to be corrected. The LR particles and the HR particles were regrouped on a spherical grid with grid angle of 2 degrees, resulting in 669,741 LR and 602,289 HR grid particles. The 2 degree grids assure a highly accurate representation of the large scale tidal fields (Stoehr 1999) although they

represent only 2.2 % of the total particle number in our highest resolution simulation.

As a first project we had studied the feasibility of the resimulation of a pair of galaxies having properties similar to the Local Group: a separation of about 725 kpc, a radial velocity of -70 km/s and masses of about 10^{12} $M_\odot$ per galaxy. There are only very few candidates in the HR region of M3 being "close" to these values. The values of the best pair in M3 are shown in Table 3.3. There was no approaching pair, i.e. a pair having a negative relative velocity. A series of test runs showed that the properties of the original pair could not be reproduced with sufficient accuracy. The distance of the resimulated pair as well as the relative velocity were too large in all of our test simulations. It turned out that a high force and integration accuracy at high redshift was indispensable for a reproduction of these quantities. Table 3.3 shows the values for a low and a high accuracy run. For the latter the parameters were chosen according to Power et al. (2003).

However, even with this setting the results of the best pair were not satisfactory for our needs and we decided to resimulate a single galaxy, i.e. candidate G-2 from Table 3.1. In addition to an easier resimulation, this choice allowed for a much higher resolution in the selected object because the HR region could be much smaller. Moreover, we could choose among more good candidates, allowing us to put constraints on the merging history, the mass and the position within M3.

After the candidate halo has been selected a starting redshift for the resimulation had to be determined.

3.2.1.3 Starting redshift

At very high redshifts, just after the epoch of recombination, density fluctuations were so small that the evolution of the structure in the Universe was almost perfectly linear. According to linear theory, using the spherical top-hat approximation (section 1.2.6), density fluctuations reach an extrapolated density contrast of 1.686 before starting to collapse. In order to save computer time, the simulation should start at the point where the first objects - in the CDM cosmologies these are the smallest objects - just are about to start to collapse. This moment depends therefore on the mass resolution of the simulation.

The criterion implemented in ZIC is the one used by Bertschinger (1995). The starting redshift of the evolution of the simulation is here determined by the largest displacement a particle obtains. All displacements are scaled to a redshift where the maximum displacement is a given fixed fraction of the mean interparticle separation. In the case of Bertschinger (1995) this fraction was 1.

3. Satellite Galaxies of the Milky Way

Pair	original (in M3)	low accuracy	high accuracy
Mass 1 [$M_\odot$]	1.8×10^{12}	1.7×10^{12}	1.7×10^{12}
Mass 2 [$M_\odot$]	3.0×10^{12}	2.7×10^{12}	3.0×10^{12}
Distance [kpc]	1356	1949	1896
Relative velocity [km/s]	+16	+172	+106

Table 3.3: Local group resimulation project. Very high integration accuracy - especially at high redshifts - was required in order to get low escape velocities. The original relative velocity could not be reproduced. There was no candidate pair that had an encounter velocity close to the one of the Milky Way and M31 in the Local Group. For the high accuracy run the parameters were set according to the suggestions of Power et al. (2003). The low accuracy run was performed with the parameter settings used for the M-series simulations (chapter 2).

3.2 The simulated Milky Way

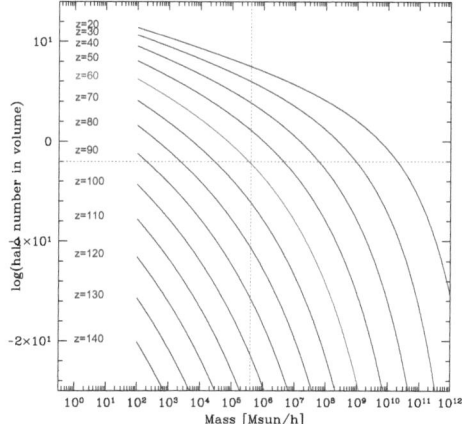

Figure 3.4: The average number of collapsed objects larger than a given mass in the comoving volume of 7.3×10^4 (Mpc/h)^3, i.e. the volume of the HR region of M3. When starting our highest resolution simulation at redshift $z = 60$, the probability that a 2-particle halo has collapsed already is smaller than 1% (dotted lines). As the simulated halo is two orders of magnitude smaller than the largest object in the HR region of M3, this starting redshift is still very conservative.

Although this procedure certainly gives safe starting redshifts, i.e. provides initial conditions that are well in the linear regime, lower redshifts closer to the actual collapse redshift of the first objects are possible. As the lower resolution simulations were not very time consuming, the starting redshift determined with the above criterion was a reasonable choice. For our highest resolution simulation however, this redshift turned out to be $z_{start} \approx 300$ corresponding to a predicted total CPU time of well over 10^5 hours for the version of GADGET at that time.

Using the mass function - here the cumulative one - determined by Sheth et al. (2001b) a better estimation of the starting redshift is possible. The cumulative mass function gives the number of collapsed objects per volume with masses larger than a given threshold mass. It is then simple to find the redshift where for example the probability that

3. Satellite Galaxies of the Milky Way

an object of the mass corresponding to the simulation particles of the highest resolution simulation has collapsed within the volume of 7.3×10^4 (Mpc/h)^3, the volume of the final uncontaminated HR region, is smaller than 1 %. Figure 3.4 shows the average number of haloes in this volume as a function of the minimum halo mass. The lines give the relations for different redshifts. The horizontal and vertical lines correspond to the minimum mass of 4×10^5 M$_\odot$/h, and the collapse probability of 1 %, respectively. These are the settings for our highest resolution simulation which we have called GA3n (see 3.4). It is clear from the plot that a starting redshift of $z = 60$, the redshift that we have used, will be sufficient for this simulation. The choice is still very conservative as the object we simulated is two orders of magnitude smaller than the two clusters that are present in the HR volume of M3. $z = 60$ is safe even for these clusters although they will start collapsing much earlier than the Milky Way sized halo.

3.2.1.4 Contamination

One of the most critical aspects of resimulations is the control of the region into which where LR particles penetrate. "Uncontaminating" a simulation is done by running the resimulation, tracing back LR particles that fall into a given region and replacing them with HR particles in the initial conditions. Usually one iteration step is sufficient to produce a LR particle-free final region. This has to be done only for the lowest resolution resimulation. The differences in the distribution of the LR particles between the runs of different resolutions are negligible.

For the G-series simulations we tracked back the particles from a sphere of 1 Mpc/h around the centre of the galaxy (with a virial radius of 205 kpc/h). With this setup 18% of the HR particles fell into the virial radius of the main halo.

3.2.1.5 Simulations

The GA-series consists of four simulations of the same galaxy (candidate G-2) with mass resolutions increasing each time by a factor of 9.33 starting from the one of the parent simulation, the ΛCDM M3 simulation, with a mass resolution of 1.6×10^8 M$_\odot$/h. Table 3.4 gives the values for the initial conditions.

For the G-series simulations we have used Volker Springel's ultra-fast code GADGET 2.0 with a timestep inversely proportional to the acceleration of the particles and the parameters given above. Note however, that GA3n has been run with a slightly older and less accurate version.

One important issue of the initial condition set up was to assure that the low frequency displacements in all four simulations are identical. Due to the nature of ZIC, the random values used to construct the displacement field change when the size of the HR or SHR

3.2 The simulated Milky Way

	$m_{particle}$ [M$_\odot$/h]	N_{HR}	N_{LR}	N_{tot}	z_{start}	ϵ [kpc/h]
GA0n	1.677×10^8	68323	1272197	1340520	70	1.4
GA1n	1.796×10^7	637966	1272382	1910348	80	0.8
GA2n	1.925×10^6	5953033	1272434	7225467	90	0.38
GA3n	2.063×10^5	55564205	1272030	56836235	60	0.18

Table 3.4: Initial condition parameters of the GA-series. Note the low starting redshift of GA3n.

grid changes. We therefore have used the same full size SHR grid and the same random realization for all simulations of this series. For the simulations with lower resolution than GA3n, the modes in k-space smaller than the Nyquist-frequency of the SHR particles were set to zero.

The position and velocity data were saved in 108 outputs from redshift $z = 37.6$ to $z = 0$. 60 outputs are equally spaced in $\log(1/(1+z))$. The other outputs were selected at lower redshifts so that the time between two outputs below $z = 2.3$ is approximately constant.

Note that the simulations used in this and the next chapter are not identical with the ones used by Stoehr et al. (2002). The simulations have been rerun with higher force accuracy with the newer version of GADGET. In order to distinct them form the previous simulation set, we have termed these simulations GA0n, GA1n, GA2n and GA3n, respectively.

3.2.1.6 Scaling

We scaled the simulated haloes down by a factor of 0.91 in length and $0.91^3 = 0.74$ in mass scale so that the peak of the circular velocity curve closely matches 220 km/s, the value of the Milky Way. The density and time scales remain unchanged.

Note that unless stated otherwise all values given below in this and in the next chapter correspond to the scaled simulations!

3. Satellite Galaxies of the Milky Way

	GA0	GA1	GA2	GA3
m_{scaled} [$M_\odot$/h]	1.24×10^7	1.33×10^7	1.42×10^6	1.71×10^5
ϵ_{scaled} [kpc/h]	1.2	0.72	0.34	0.17
R_{200} [kpc/h]	196.1	194.4	194.8	195.5
M_{200} [$M_\odot$/h]	1.75×10^{12}	1.70×10^{12}	1.71×10^{12}	1.73×10^{12}
N_{200}	14097	128276	1204411	10089396
N_ϵ	9	21	52	43
c	9.25	9.18	9.94	9.97

Table 3.5: Values of resimulated haloes. Here, N_{200} is the number of DM particles within R_{200}, N_ϵ the number of particles that were closer than one softening length to the centre of the halo and c the NFW concentration parameter (see section 2.4.2).

3.2.2 Simulation results

We ran the full post processing pipeline (halo-finding, property-finding, profile-finding, merging-history finding, contamination-finding) over the simulations including SUBFIND to identify self bound substructure within the main halo of the Milky Way. Figure 3.8 shows a cubic region of 570 comoving kpc/h on a side around the haloes at $z = 0$ for the four simulations. Figure 3.9 shows the assembly of the halo in GA3n. The images correspond to the redshifts $z = 10, 5, 2$ and and 1. The properties of the main halo and the subhaloes are discussed in the following sections.

3.2.2.1 Density profile and circular velocity curve

The determination of the density profile of DM haloes has been and still is an important challenge in cosmology. Especially the value of the inner slope of the profile is uncertain. However, it can play a crucial role for example for the possible annihilation of DM particles and their detectability (see chapter 4). Many authors have tried to determine the inner slope of dark halo density profiles through physically based analytic arguments (Peebles 1980; Hernquist 1990; Syer & White 1998; Nusser & Sheth 1999; Subramanian et al. 2000; Dekel et al. 2002) but despite very interesting ideas the final answer is still missing. Numerical simulations seem to be an ideal tool to solve the problem.

As already mentioned, Navarro, Frenk & White (1995a) performed high-resolution simulations and fitted the "universal" density profile

$$\rho_{NFW}(r) = \frac{\delta_c}{\frac{r}{r_s}(1+\frac{r}{r_s})^2}. \quad (3.1)$$

to their simulated haloes. Here r_s is a characteristic length and δ_c a characteristic density. At the scale radius r_s the slope changes from $\propto r^{-1}$ to $\propto r^{-3}$.

In subsequent papers, NFW extensively studied halo profiles for different underlying cosmologies as well as for different ranges of halo masses and compared their results to observed galaxy circular velocity curves and to profiles of X-ray clusters (Navarro, Frenk & White (1995b)). They found the profile to be a good description for the simulated haloes in all of the studied cosmologies over four orders of magnitude in mass and over two orders of magnitude in the distance from the halo centre.

In addition, they give a correlation between the formation time of the halo and its characteristic density (Navarro, Frenk & White (1996)). It then is possible to express the NFW profiles for a given mass, or equally for a given radius r_{200}, by a single parameter: the concentration parameter $c = r_{200}/r_s$. The density profile writes as

$$\rho_{NFW}(r) = \frac{3H_0^2}{8\pi G}(1+z)^3\frac{\Omega_0}{\Omega(z)}\frac{\delta_c}{cx(1+cx)^2} \quad (3.2)$$

3. Satellite Galaxies of the Milky Way

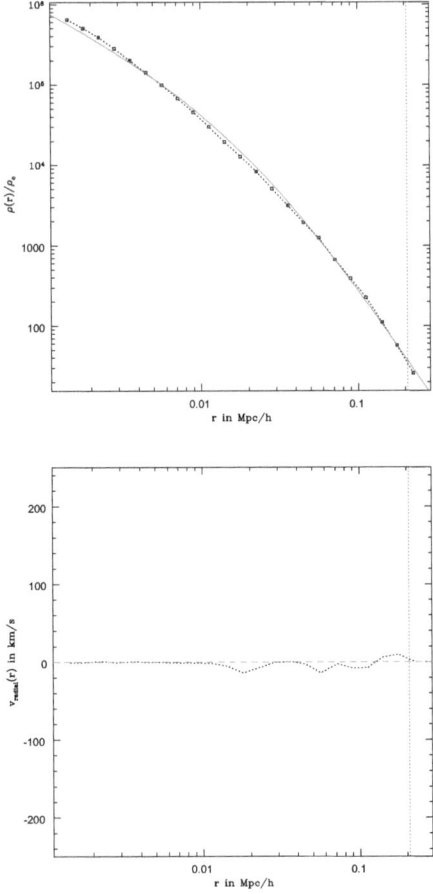

Figure 3.5: *Upper panel*: density profile of the GA3n main halo in units of the critical density. The solid line shows an NFW profile with concentration parameter $c = 9.79$. The vertical line indicates the virial radius. Down to about 1% of the virial radius, the NFW profile provides an excellent fit to the simulation data. *Lower panel*: radial velocity within the virial radius. For easier comparison with the corresponding plots for the G-2 candidate, these two and the next plot were not scaled to match the MW peak rotation velocity.

3.2 The simulated Milky Way

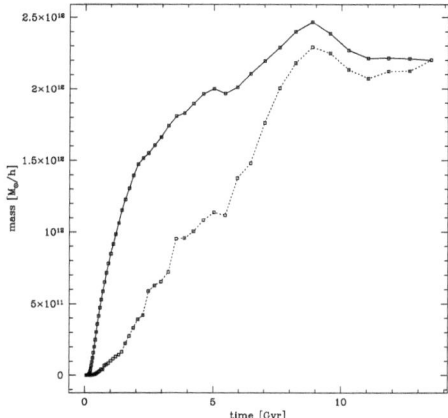

Figure 3.6: Mass growth history of the GA3n main halo. The solid line shows the mass of all progenitors of the final halo, the dotted line shows the mass of the most massive progenitor.

3. Satellite Galaxies of the Milky Way

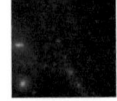

Figure 3.7: Merging history of GA3n. Only subhaloes with more than 8090 particles are taken into account. This corresponds to the 10 particle limit in M3.

Figure 3.8: Density distribution in logarithmic scale of the $z = 0$ snapshot of the simulations GA3n, GA2n, GA1n and GA0n. The box side length corresponds to three times the virial radius (i.e. 570 kpc/h).

3. Satellite Galaxies of the Milky Way

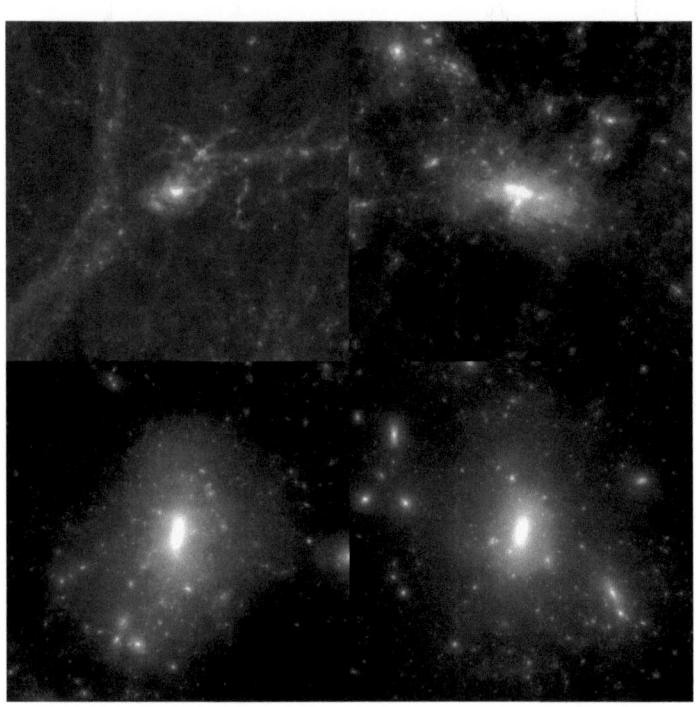

Figure 3.9: History of the GA3n halo assembly. The images correspond to the redshifts $z = 10, 5, 2$ and and 1.

3.2 The simulated Milky Way

with $x = r/r_{200}$ at the redshift z, and the characteristic density

$$\delta_c = \frac{200}{3} \frac{c^3}{\ln(1+c) - c/(1+c)}. \tag{3.3}$$

The radius r_{200} is linked to the mass M_{200} through

$$r_{200} = 1.63\ 10^{-2} \left(\frac{M_{200}}{M_\odot}\right)^{1/3} \left(\frac{\Omega_0}{\Omega(z)}\right)^{-1/3} (1+z)^{-1} h^{-2/3} \text{ kpc}. \tag{3.4}$$

The circular velocity for this model is

$$V_{200} = \left(\frac{G\ M_{200}}{r_{200}}\right)^{1/2} = \left(\frac{r_{200}}{\text{kpc}}\right) \left(\frac{\Omega_0}{\Omega(z)}\right)^{1/2} (1+z)^{3/2} h^{-1} \text{ km s}^{-1} \tag{3.5}$$

and

$$V_{circ}(r) = V_{200} \sqrt{\frac{1}{x} \frac{\ln(1+cx) - (cx)/(1+cx)}{\ln(1+c) - c/(1+c)}} \tag{3.6}$$

Two parameters fully specify the profile. This can be either (r_s, δ_c) or (r_{200}, c) or (r_{max}, v_{max}). Where the latter pair refers to the maximum of the circular velocity curve $V_{circ}(r)$.

By numerically solving equation 3.3 for the concentration parameter c, the first pair of parameters can easily be converted into the second one whereas the reverse conversion is trivial.

At the maximum of the circular velocity curve cx has to fulfil the relation

$$\ln(1 + c\ x_{max}) = \frac{c\ x_{max}}{1 + c\ x_{max}} + \frac{c^2\ x_{max}^2}{(1 + c\ x_{max})^2} \tag{3.7}$$

We solve this equation numerically and find that

$$c\ x_{max} \approx 2.1625814 \tag{3.8}$$

and therefore

$$r_{max} \approx \frac{2.1625814\ r_{200}}{c} = 2.1625814\ r_s \tag{3.9}$$

This together with equations 3.5 and 3.6 allows to convert from (r_{200}, c) to (r_{max}, v_{max}). For the inverse conversion, the equation

$$\left(\frac{v_{max}}{\text{km s}^{-1}}\right) \left(\frac{\text{kpc}}{r_{max}}\right) 21.62996674 \left(\frac{\Omega_0}{\Omega(z)}\right)^{-1/2} (1+z)^{-3/2} h = \tag{3.10}$$

$$\sqrt{\frac{c^3}{\ln(1+c) - c/(1+c)}} \tag{3.11}$$

has to be solved for c and the result inserted in equation 3.7. In what follows in this chapter we will often use the pair (r_{max}, v_{max}) which can be extracted very robustly from the circular velocity curve of equation 2.3 by just sorting the particles according to their distance, counting them and inserting the result in equation 2.3. The centre of the halo is defined as the position of the most bound particle of the halo. In addition, using the circular velocity curve for the profile fitting instead of the density profile has the advantage that no binning is necessary and that therefore the curve can be plotted down to the very centre of the halo.

This technique allows therefore to give NFW "fits" and NFW-mass estimates (using equation 3.6) for haloes with very few particles.

Tormen, Bouchet & White (1997) ran simulations and confirmed the findings of NFW. Moore et al. (1999) and Ghigna et al. (2000) preferred to fit their haloes to a fitting formula with an inner slope of -1.5 instead of -1. Recently Power et al. (2003) performed and extensive parameter study using two different simulation codes: GADGET and PKD-GRAV and found good agreement, with the profiles to be fit better by the NFW profile than by the one proposed by Moore et al. (1998). This is interesting as these authors had used PKDGRAV to find their inner slope. Power et al. (2003) however point out that the slope at their innermost resolved point is not yet exactly equal to -1 but is a bit higher: around 1.2-1.4 depending on the particular halos considered.

As this brief overview shows, the determination of the inner slope of the density profile of DM haloes is far from a simple task. The first difficulty is the required mass resolution that is only achievable using some variant of the resimulation technique. But even using this method, with each increase of the mass resolution of one order of magnitude, only a factor of $\sqrt[3]{10} \approx 2.15$ in spatial resolution is gained if the softening ϵ is set proportional to the mean particle separation of the SHR particles. Using the criterion derived from Power et al. (2003) the softening and therefore the spatial resolution scales as $\sqrt{N_{200}}$ i.e. ≈ 3.16 for one order of mass resolution increase.

A safe determination of the inner slope of the density profile, or of an other fit if it turns out that the slope is slowly varying, would be possible with a spatial resolution increase of an other one of two orders of magnitude. Even with planned computing and data processing facilities this will be a challenging task. With a convergence study however, some extrapolation beyond the measurement of the highest resolution simulation is possible.

The difficulty is to achieve sufficient numerical accuracy in the very centre of the DM halo. Even the smallest statistically significant bias in the integration of the innermost particles can lead to an energy increase or decrease and thus expulsion or accumulation of particles from or to the centre resulting in a lower or higher density, respectively. This integration however is by construction not perfect. It is done with approximated

forces (stemming from the tree evaluation as well as from the softened potential) and time steps that are adaptive and chosen according to the properties of the simulation particle or the properties of its local environment (the acceleration of the particle or the local velocity dispersion for example).

Power et al. (2003) performed extensive tests and came up with parameters that they considered "safe". We chose even more conservative parameters than proposed by the guidelines of these authors.

Figure 4.2 in the next chapter shows the circular velocity profiles for our four resimulated haloes together with an NFW profile with parameters $r_{200} = 270$ kpc and $c = 10$. We plot these curves down to a distance from the centre equal to the softening length.

The profiles of the simulations agree well and the NFW profile provides a reasonable fit. We overplotted also a fit of the parabolic formula we propose below. Especially at distance of about 1-5 kpc from the halo centre, this profile matches the data better than the NFW profile. Note, however, that this profile has an additional free parameter, unlike the NFW profile which is fixed shape once the peak of the circular velocity curve is determined.

Whether this new profile (Stoehr et al. (2002), hereafter SWTS) is a better description than the NFW profile for the very inner part of velocity profiles of DM haloes in general is unclear to date. Unfortunately, the simulations that were available to us - other than the GA-series - have been run with versions of GADGET that did not faithfully reproduce the inner 1% of the DM haloes so that the question could not be addressed.

An interesting side effect of the SWTS profile is that is predicts a "core". Some observations currently favour DM haloes with cores (McGaugh & de Blok 1998; de Blok et al. 2001b). These issues are discussed in more detail in section 4.4.

3.2.2.2 Bound vs. unbound subhalo particles

For the following analysis we identify the subhaloes using the SUBFIND algorithm (Springel et al. (2001a), see also section 2.5.1).

As the matter distribution of the main halo on the scale of the subhalo can be considered homogenous, the unbound particles within the volume of the subhalo and are not dynamically relevant for the motion of the stars in the substructure potential.

We note that in the article of Stoehr et al. (2002) *all* particles within the volume of the subhalo were used to estimate the subhalo circular velocity curve. This was conservative for our purpose. The difference of the values of r_{max} and v_{max}, characterising the subhaloes, however is rather small whether one takes the unbound particles into account or not. This is due to the fact that subhaloes are much denser than the DM of the surrounding main halo, resulting in relatively few unbound particles within the volume

3. Satellite Galaxies of the Milky Way

of the substructure.

Figure 3.10 shows the r_{max} and v_{max}-values of the substructure circular velocity curves with and without the contribution of the unbound particles for the simulations GA1n, GA2n and GA3n. The two methods give remarkably similar v_{max}-values. Some of the r_{max} values differ by up to a factor of two (in one extreme case by a factor of 4).

The total mass bound in subhaloes is 1.7%, 3.0%, 5.1% and 4.1% of the mass of the main halo in GA0n, GA1n, GA2n and GA3n, respectively. The fluctuations in this number are due to the fact that the most massive subhaloes – which dominate the total substructure mass (see section 4.5.1) – are subject to Poisson statistics. Although the resimulation in principle should give exactly the same results for each run, this is not true in reality. For the mass scales corresponding to the subhaloes of a Milky Way halo, the evolution of the density field is highly nonlinear at lower redshifts. Tiny deviations in the initial conditions and slightly different forces can therefore lead to the situation where a subhalo has already been accreted at $z = 0$ in one simulation where in an other it has not yet merged with the main halo.

The above values are for the mass fraction in substructure are in agreement with recent studies of image flux ratios in multiply lensed quasars (Chiba 2002; Dalal & Kochanek 2002). These authors argue that their findings are a strong support for the CDM model.

The most massive subhaloes have 1.8%, 1.9%, 2.3% and 0.8% of the virial mass in the simulations GA0n, GA1n, GA2n and GA3n, respectively. For GA3, this corresponds to 88832 DM particles.

3.2.2.3 Backtracking

We can have a closer look on the change of the dynamic structure of the subhaloes by tracing them back to the redshift just before they merged with the main halo and compare the structural parameters r_{max} and v_{max} as well as the masses pre and post infall. We restricted to subhaloes with more than 100 particles.

The distribution of the infall-redshifts for GA3n is shown in the upper panel of Figure 3.11. The lower panel shows the cumulative distribution of these redshifts normalised to the total subhalo number, i.e. 412. It is striking that most of the subhaloes that can still be detected today were accreted only very recently. Half of the subhaloes were accreted later than $z \approx 0.4$. Only 10% of the haloes were accreted before $z \approx 1.6$; only 1% before $z \approx 3.8$. The oldest remnant present in GA3n at $z = 0$ was accreted at $z \approx 7.2$. All this implies that tidal disruption is very strong which is in agreement with results from Gao Liang (2003) (private communication). The outer regions of the infalling haloes get stripped away and the material is lost to the halo of the MW.

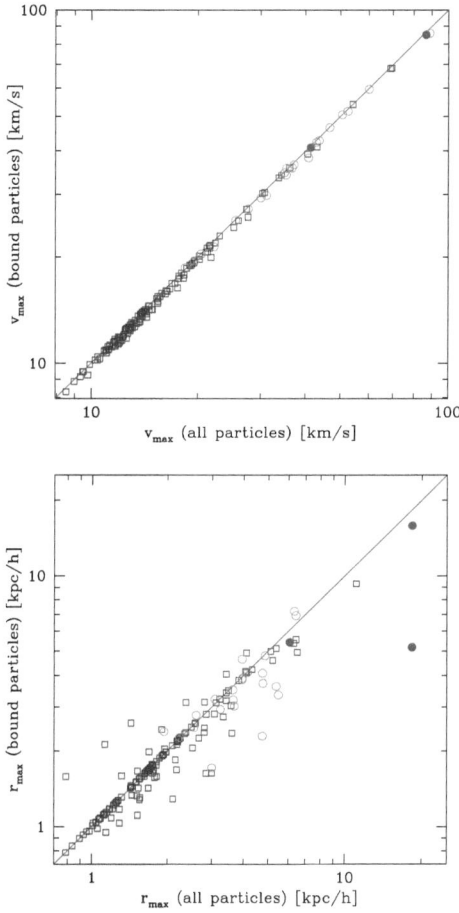

Figure 3.10: *Upper panel:* v_{max}-values for the subhaloes with more than 300 particles in the simulations GA1n (dots), GA2n (open circles) and GA3n (open squares). The values have been measured once taking into account th gravitationally bound particles only, once all simulation particles. *Lower panel:* as above for the r_{max}-values.

3. Satellite Galaxies of the Milky Way

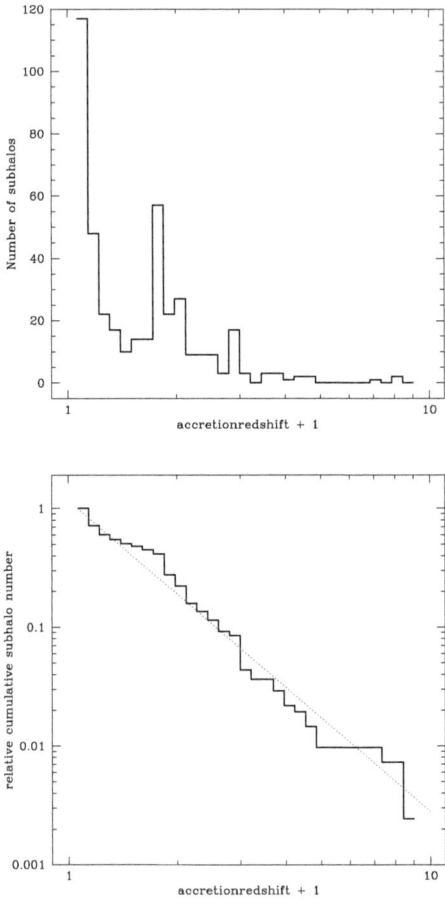

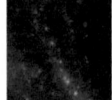

Figure 3.11: Accretion of the subhaloes with more than 100 particles. The upper panel shows the number of accreted subhaloes that survived to $z = 0$ as a function of their accretion redshift. The lower panel shows the cumulative normalised survival fraction. We find that this fraction can be approximated by $(1 + z)^{-2.65}$.

3.2 The simulated Milky Way

The fit (dotted line) corresponds to

$$f(z) \approx (1+z)^{-2.65}. \tag{3.12}$$

The line in the plot is shifted to match the value of 100% at $z = 0.06$ which is the redshift of our penultimate output. $f(z)$ is the relative number of surviving haloes which are accreted onto the main halo before z.

However, not only the outer regions of the haloes change upon infall, also their internal structure is altered. The change in the circular velocity profiles is discussed in the next section.

In this section we study the changes on an object-by-object basis for the subhaloes of one simulation. Figure 3.12 shows the r_{max} and v_{max} values before the infall and at redshift $z = 0$ for all four simulations. For all subhaloes the peak rotation velocity decreases and for most of them the position of the peak moves inwards. This change is not a step function at the redshift of accretion. It gradually increases with the time the subhalo is orbiting in the main halo. This can be seen from Figure 3.13. Although the scatter is huge, the average relative change in r_{max} or v_{max} is a rather smooth function of the accretion redshift.

Figure 3.14 shows the mass-loss that the subhaloes suffer upon infall into the main halo as a function of the distance from the centre of the main halo at redshift $z = 0$. In the upper panel, the mass of the pre-infall halo is the FOF mass whereas for the post-infall mass the SUBFIND mass is used. As expected, all sub haloes suffer from tidal stripping. This effect is stronger in regions of higher density. It is interesting to observe, that this trend is clearly visible (solid histogram line for GA3n subhaloes) although only a snapshot of the subhalo distribution is analysed. The fact that this trend is observable in a snapshot means that the disruption-time of a subhalo - or the time where significant mass loss is occurring -, is comparable to the time it takes a subhalo to cross the main halo.

SUBFIND determines the mass of the subhaloes by finding the particles that are gravitationally bound to the substructure. This naturally results in smaller masses for subhaloes that are closer to the centre of the main halo being deeper in the potential well. In order to assure that the observed effect is neither due to the different algorithms used for the mass determination (FOF for the pre-infall haloes and SUBFIND for the subhaloes) nor to the SUBFIND itself, we redid the above analysis, this time only using the structural parameters r_{max} and v_{max}. Assuming that the pre-infall haloes and the subhaloes have NFW density profiles we can compute masses "m_{200}" for all of them and determine the mass loss as before. Some haloes seem to have gained mass which is an artefact from the approximation with the NFW profile. The result of this test is shown in the lower panel of Figure 3.14. It confirms our previous findings.

3. Satellite Galaxies of the Milky Way

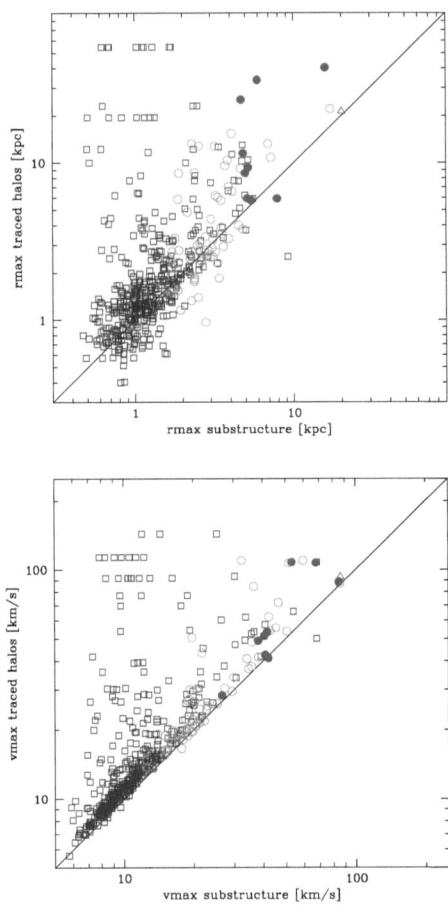

Figure 3.12: Change of r_{max} and v_{max} values of infalling subhaloes.

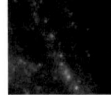

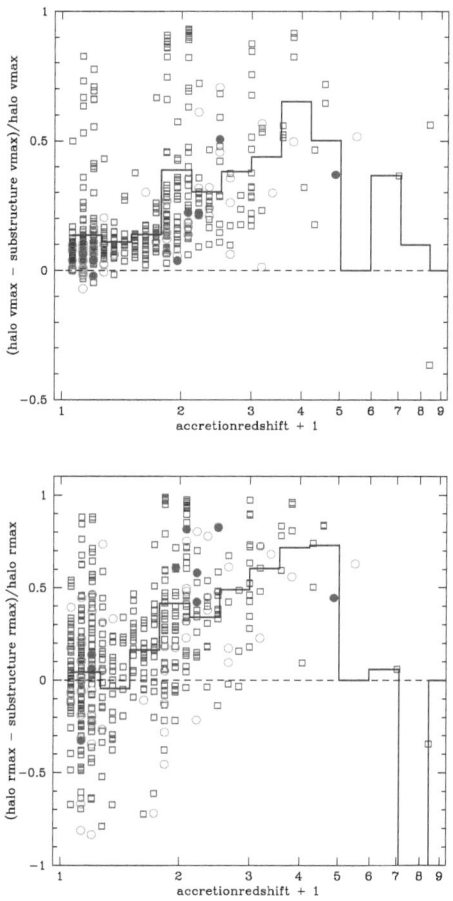

Figure 3.13: Relative change of v_{max} and r_{max} of the haloes upon infall into the main object as a function of its accretion redshift: The peaks of the subhalo circular velocity curves are gradually decreasing and moving inwards.

3. Satellite Galaxies of the Milky Way

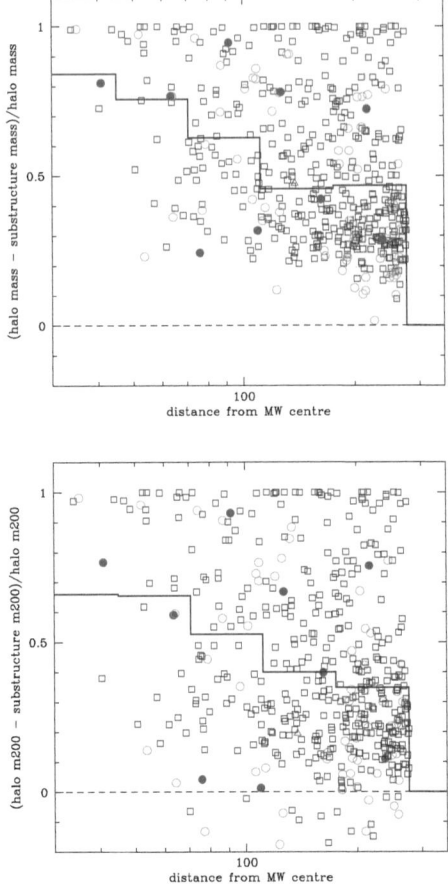

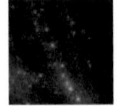

Figure 3.14: Mass-loss fraction of haloes due to their infall into the main halo as a function of their distance to the MW centre at $z = 0$. Although only one "snapshot" of the evolution is analysed, the trend is clearly visible. The upper panel shows the mass-loss computed directly with the FOF and SUBFIND masses, whereas for the plot in the lower panel only the (r_{max}, v_{max}) values have been used and an NFW profile has been assumed.

3.2 The simulated Milky Way

The resolution of the simulations, especially GA3n, is so high, that the orbits of the subhaloes in the Milky Way potential can be followed up to high redshifts. An interesting follow-up project would be to determine the full path of the subhalo and track the evolution of structural parameters, correlating them with the initial orbital parameters and the initial mass. This together with the infall distribution would make it possible to better understand the buildup of DM haloes and would probably allow improvement of semi-analytic modelling techniques (Kauffmann, White & Guiderdoni (1993);Cole et al. (2000); Taylor & Silk (2003)).

3.2.2.4 Circular velocity curves

In this subsection we compute the circular velocity curves of subhaloes and fit analytical profiles to them that can be used to predict the central velocity dispersions of observed stellar distributions. Only a few such studies exist to date. Moore et al. (1999); Ghigna et al. (2000) obtained profile fits from simulations like the GA-series. They found their haloes to follow steep slopes of -1 to -1.5.

Figure 3.15 shows the circular velocity curves of a selection of the most massive subhaloes of GA3n. As mentioned we used only the gravitationally bound particles in equation 2.3. The curves rise from the centre, peak at a distance of a few kpc and then fall rapidly. The curves are *narrower* than the NFW profile (the dashed curve of the peak for the circular velocity curve of the most massive subhalo). The corresponding density profiles are thus *shallower* than the NFW density profile, in contrast to the findings of Moore et al. (1999); Ghigna et al. (2000).

Our result however is consistent with the recent study of Hayashi et al. (2003). These authors followed a different strategy: They studied the change of the density circular and velocity profiles of dwarf galaxy DM haloes – set up initially with NFW profiles – as they fell into a fixed NFW potential representing the Milky Way.

We fitted the circular velocity curves of our subhaloes with parabolae:

$$\ln\left(\frac{V_{circ}(r)}{V_{max}}\right) = -a \, \ln\left(\frac{r}{r_{max}}\right)^2 \qquad (3.13)$$

The stars in the observed Milky Way dwarfs extend to a few kpc typically. We therefore adjusted the values of a so that equation 3.13 is a good fit to the data for $r < r_{max}$. In this range the profiles are remarkably well fit by our formula. This is however misleading. Extending the parabolic fit beyond (see also section 4.8)

$$r_{core} = r_{max} \, e^{-\frac{1}{2a}} \qquad (3.14)$$

would lead to a decline of the density in the centre which is clearly unphysical. For $r < r_{core}$ we therefore assume a constant maximal density for the profile. The r_{core}

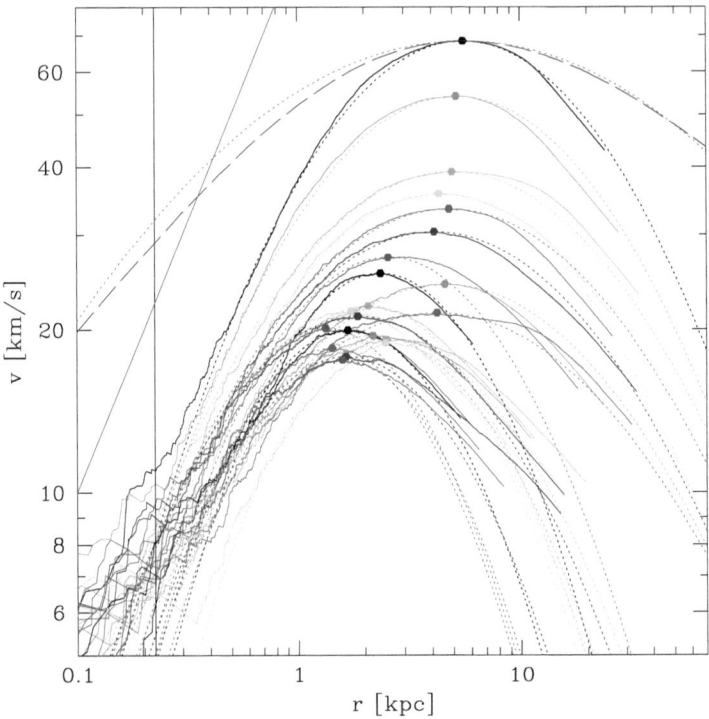

Figure 3.15: Circular velocity curves for the GA3n subhaloes ranked 1, 2, 4,... 40 in mass (solid) together with corresponding parabolic fits (dotted). For comparison an NFW profile (dashed) and a parabolic profile with $a = 0.074$ (dotted) - the value for the main halo - are overplotted on the most massive subhalo. The vertical solid line shows the softening length, the diagonal line shows the profile slope corresponding to a constant density.

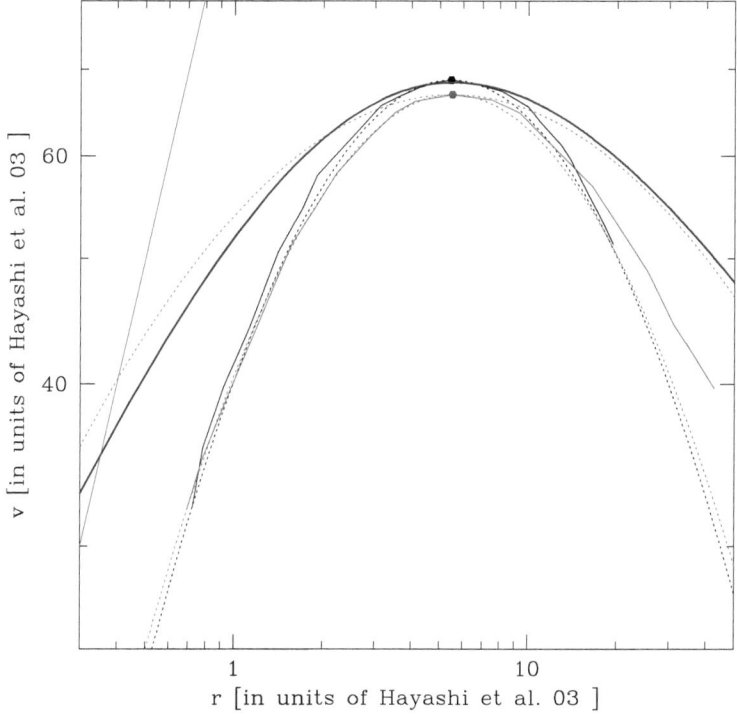

Figure 3.16: As Figure 3.15 for two representative circular velocity curves taken from Hayashi et al. (2003). The corresponding a values are 0.16 (upper curve) and 0.18 (lower curve), respectively.

3. Satellite Galaxies of the Milky Way

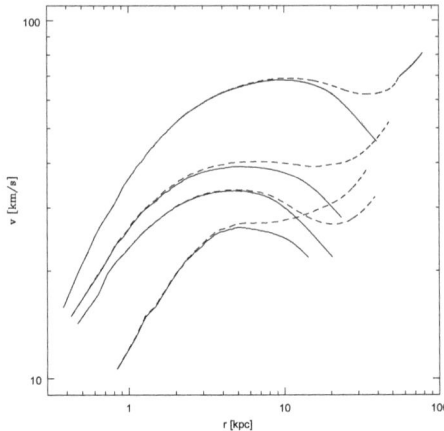

Figure 3.17: Subhalo circular velocity curves for the GA3n subhaloes 1, 4, 7 and 10. The solid lines show the profiles taking only the bound particles into account. Dashed lines correspond to profiles using all particles within r of the subhalo centre.

values are too small for equation 3.14 to be relevant in this chapter. It will be used however in the next chapter.

Note that the definition of equation 3.13 differs from equation 1 of Stoehr et al. (2002). To allow for simpler expressions in section 4.8 we changed to the natural logarithm. This results in

$$a_{thesis} = \frac{a_{Stoehr\ et\ al\ 2002}}{\ln(10)} \approx \frac{a_{Stoehr\ et\ al\ 2002}}{2.3} \qquad (3.15)$$

Especially for $r < r_{max}$ equation 3.13 provides excellent fits to *all* of the substructure profiles in Figure 3.15 as well as for the halo profiles computed by Hayashi et al. (2003) as shown in Figure 3.16. In the outer regions the profiles start to be under-predicted where the potential gradient of the main halo at the position of the particles starts to become comparable to that of the subhalo. The particles then start to be bound to the main halo. This can be seen from Figure 3.17 which shows for the GA3n subhaloes 1, 4, 7 and 10 the circular velocity curves including all particles (dashed) and including only the particles that are gravitationally bound to the subhalo (solid).

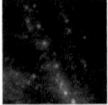

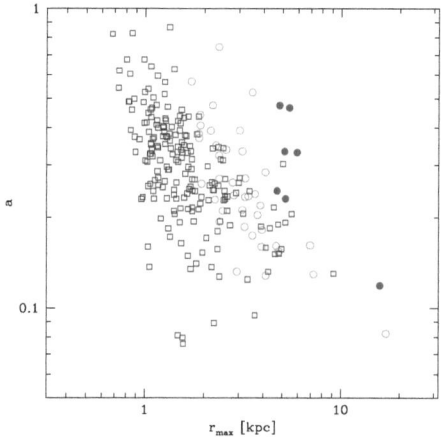

Figure 3.18: a-values for the GA1n (filled circles), GA2n (open circles) and GA3n (open squares) subhaloes with more than 300 particles.

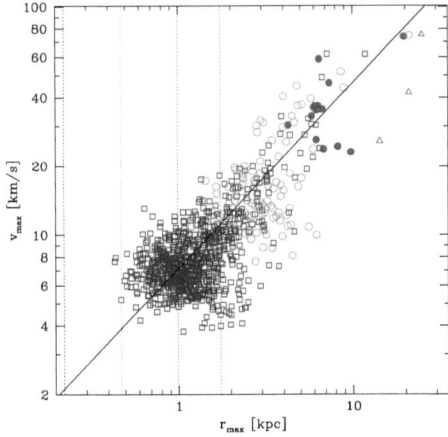

Figure 3.19: (r_{max},v_{max}) values for the four simulations (symbols as above). The one-third most massive subhaloes are shown, i.e. 3 of GA0n (open triangles), 13 of GA1n, 102 of GA2n and 780 of GA3n. The vertical lines correspond to the softening lengths of the simulations, the diagonal line shows the fit 3.16

Figure 3.18 shows the values a for all of the subhaloes containing more than 300 particles in the simulations GA1n, GA2n and GA3n. There is a slight trend to smaller r_{max} values for simulations with higher mass resolution.

Figure 3.19 shows the structural parameters (r_{max}, v_{max}) for the one third most massive subhaloes in GA0n, GA1n, GA2n and GA3n, respectively. The vertical lines indicate the softening lengths of the simulations. The solid diagonal line corresponds to the fit

$$v_{max} = 7.08 \left(\frac{r_{max}}{\text{kpc}}\right)^{0.81} \text{ km/s} \qquad (3.16)$$

3.2.2.5 Object-by-object comparison

The fact that the initial conditions of the GA-series were set up with identical small scale displacement fields allows us to compare the subhaloes between simulations of different resolution on a object-by-object basis.

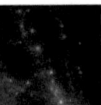

In Figure 3.20 we show the (r_{max}, v_{max}) values of the 10 most massive subhaloes in GA2n that have counterparts subhaloes within the virial radii of GA3n. We find that despite the highly non-linear nature of the system at low redshifts the agreement between the values in the two different simulations is remarkably good. For all subhaloes the a values are rather faithfully reproduced. There is a slight trend to larger r_{max} and smaller v_{max} values for GA3n subhaloes. This agreement is even more striking as they were run with different versions of GADGET.

3.3 Observations

Although the satellites within the Milky Way halo are – per definition – closer than about 270 kpc to the Sun, they are difficult to observe. Mateo (1998) shows that compared to the assumed homogenous distribution over the sky, the dwarfs have been found preferentially at high galactic latitudes where the contamination from the Milky Way disk is small. This difficulty is also expressed by the fact that the last two dwarf galaxies were only detected rather recently (Irwin et al. 1990; Ibata et al. 1994). The observed number of 11 satellites of the Milky Way therefore is probably an underestimate of the actual number.

The two largest satellites are the Magellanic clouds. The observed peaks of the circular velocity curves are about 60 km/s at 2.5 kpc and about 50 km/s at 5 kpc for the SMC and the LMC, respectively (Stanimirovic 2000; van der Marel et al. 2002). They are roughly similar to the two most massive haloes in our simulation. Besides the two Magellanic clouds, all 9 other satellites are dwarf spheroidals. The closest satellite is Sagittarius at 24 kpc from the Sun.

3.3.1 Milky Way dwarf spheroidals

The dwarf spheroidals contain no interstellar medium that could be used to determine their rotation velocity curves. In addition, they have very low surface brightnesses so that their kinematic structure can only be determined using radial velocities of individual stars (Mateo 1994; Olszewski 1998). Additional observational difficulties are contamination by sky emission lines, and the need to faithfully reject stars that are not members of the dwarf galaxy.

King (1962) found that the radial distribution of the stars in a globular cluster can be well fitted by

$$\rho(r) = \frac{k}{\pi r_c [1 + (r_c/r_t)^2]^{3/2}} \frac{\left[\arccos(z)/z - (1-z^2)^{1/2}\right]}{z^2}, \tag{3.17}$$

3. Satellite Galaxies of the Milky Way

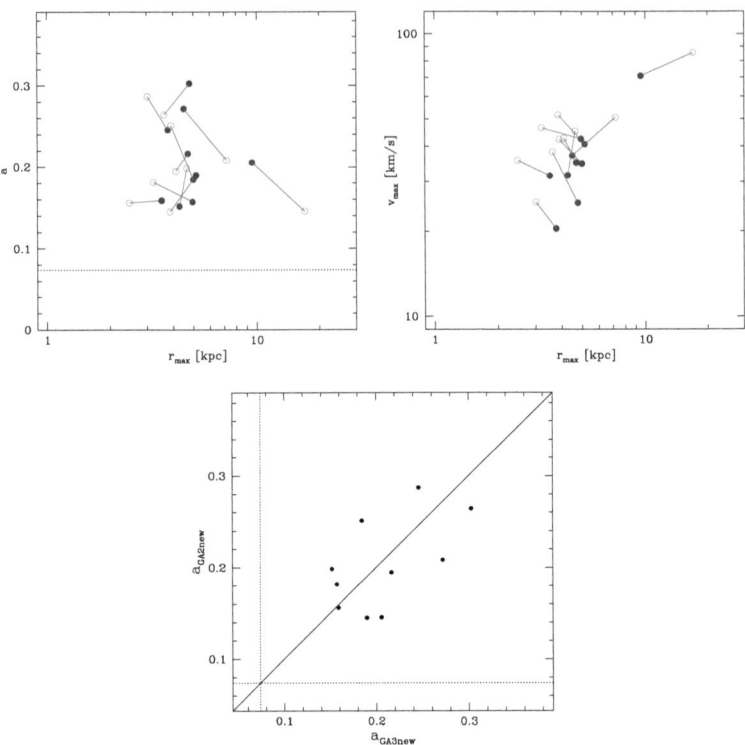

Figure 3.20: *Upper left panel:* r_{max} and a values for matching substructure pairs in GA2n (open circles) and GA3n (dots). Although there is some scatter, for some pairs the matching is very good. The horizontal line corresponds to the a-value of the main halo. *Upper right panel:* The correlation between r_{max} and v_{max}. There is a significant trend that GA3n subhaloes have lower peak velocities which is due to the fact that GA3n has been run with an older version of GADGET. *Lower panel:* a-values of matching pairs. The agreement is quite reasonable. The dashed lines show the main halo value.

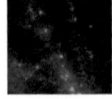

3.3 Observations

	r_c [kpc]	r_t/r_c	σ_0 [$\frac{km}{s}$]	d [kpc]
Sagittarius	0.44	6.8	11.4(19)	24
Fornax	0.46	5.1	10.5	138
Leo I	0.215	3.8	8.8	250
Sculptor	0.11	13	6.6	79
Leo II	0.16	3	6.7	205
Sextans	0.335	9.6	6.6	86
Carina	0.21	3.3	6.8	101
Ursa Minor	0.20	3.2	9.3	66
Draco	0.18	5.2	9.5	82

Table 3.6: Core radii r_c, tidal to core radius ratios r_t/r_c and central velocity dispersions σ_0 for the Milky Way dwarf spheroidals. The value in parenthesis corresponds to the *pre-disruption* model of Helmi & White (2001)

where
$$z^2 = \frac{1 + r^2/r_c^2}{1 + r_t^2/r_c^2}. \tag{3.18}$$

Here, k is a number factor, r_c the core radius and r_t the tidal radius. The core radius is the radius at which the surface brightness has fallen to half of central value, the tidal radius corresponds to the maximal extend of the stellar distribution. This profile also provides good fits to the dwarf spheroidals. Table 3.6 gives values of r_c and r_t as well as central velocity dispersions and distances from the Sun d for all the 9 known dwarf spheroidals within the virial radius of the Milky Way, taken from Mateo (1998). (The parameters k are not necessary for our analysis.)

3.3.2 Fornax and Draco

Only for Fornax (Mateo 1997) and Draco (Kleyna et al. 2001) have enough stars been observed so that radial binning and therefore the determination of the velocity dispersion profile is possible. High resolution spectra for 215 and 159 member giant branch stars had been observed for Fornax and Draco, respectively. Their radial velocities have been obtained from the Doppler shift of the star's spectral lines.

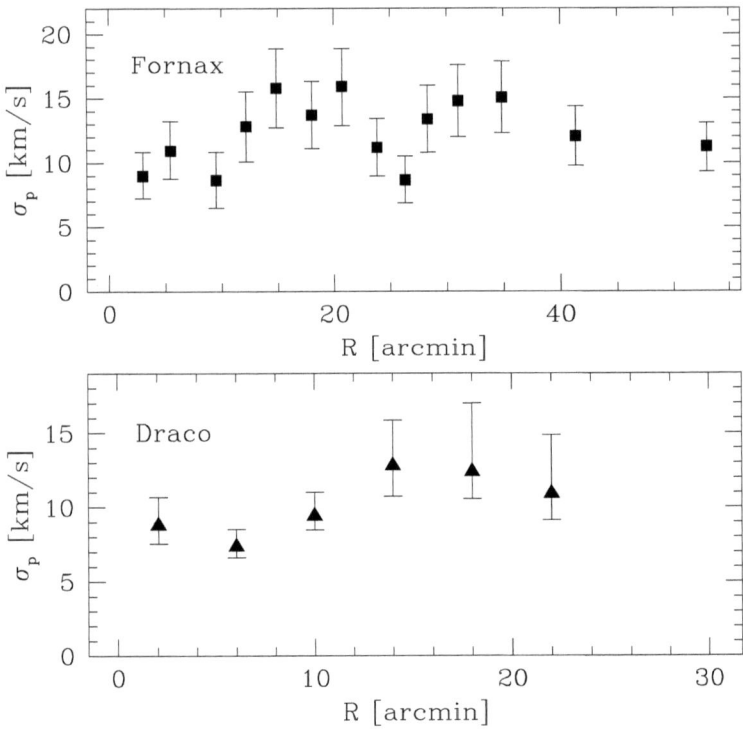

Figure 3.21: Velocity dispersion profiles for Fornax and Draco taken from Mateo (1997) and Kleyna et al. (2001), respectively.

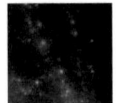

Figure 3.21 shows the corresponding projected one dimensional velocity dispersion profiles. We compare predicted velocity dispersion profiles for these systems with the full observed profiles in section 3.5.2.

3.4 Velocity dispersion

In this section we provide the formalism for the prediction of the central velocity dispersions or velocity dispersion profiles for a given distribution of stars in a spherical, time-independent potential.

3.4.1 Phase space density

Describing a large system of gravitating point masses by two-body interactions, i.e.

$$\vec{F}_i = Gm_i \sum_{i \neq j} \frac{m_j \vec{x}}{|\vec{x}_i - \vec{x}_j|^3} \tag{3.19}$$

is cumbersome. A statistical treatment of the mass distribution as a whole is possible when the system is non-collisional (see Binney & Tremaine (1987)). The distribution function $f(\vec{x}, \vec{v}, t)$, defined as the number density of point masses in phase space, fulfils a continuity equation as the total number of point-masses is conserved

$$\frac{\partial f}{\partial t} + \sum_{\alpha=1}^{6} \frac{\partial (f \dot{w}_\alpha)}{\partial w_\alpha} = 0 \tag{3.20}$$

where

$$\vec{w} = (\vec{x}, \vec{v}) = (x_1, x_2, x_3, v_1, v_2, v_3) \tag{3.21}$$

and

$$\dot{w}_\alpha = (\dot{\vec{x}}, \dot{\vec{v}}) = (v_1, v_2, v_3, -\nabla_1 \Phi, -\nabla_2 \Phi, -\nabla_3 \Phi) \tag{3.22}$$

For the special case where the force is only gravitation (in equation 3.22) the continuity equation reduces to the *collisionless Boltzmann equation*

$$\frac{\partial f}{\partial t} + \sum_{\alpha=1}^{6} \dot{w}_\alpha \frac{\partial f}{\partial w_\alpha} = 0 \tag{3.23}$$

because neither the velocities depend on the positions nor the gradient depends on the velocities (Binney & Tremaine 1987).

3. Satellite Galaxies of the Milky Way

3.4.2 The Jeans equation

Although the whole system can be described with only seven variables now, for most systems the solution is impossible. If we are interested only in mean quantities, like the mean density and the mean velocity of the point masses

$$\rho = \int f d\vec{v} \tag{3.24}$$

and

$$\bar{v}_i = \frac{1}{\rho}\int f\, v_i\, d\vec{v} \tag{3.25}$$

we can integrate equation 3.23 over all velocities. With the definition of the radial *velocity dispersion*

$$\sigma_r = \overline{v_r^2} \tag{3.26}$$

the integrated collisional Boltzmann equation in spherical coordinates for a system in a steady state can be transformed into the *Jeans equation for a spherical system* (Binney & Tremaine 1987, pages 195-198,203,279)

$$\frac{1}{\rho}\frac{d}{dr}\rho(r)\sigma_r^2(r) + 2\beta\frac{\sigma_r^2(r)}{r} + \frac{d\Phi(r)}{dr} = 0 \tag{3.27}$$

where

$$\beta = 1 - \frac{\sigma_\theta^2(r)}{\sigma_r^2(r)} \tag{3.28}$$

the coefficient of anisotropy at r. Assuming finally that the velocity distribution of the point-masses, is isotropic, then $\beta = 0$. With

$$\frac{d\Phi}{dr} = G\frac{M(r)}{r^2} = \frac{V_{circ}^2(r)}{r} \tag{3.29}$$

we get

$$\frac{d}{dr}\rho(r)\sigma_r^2(r) = \frac{\rho(r)\,V_{circ}^2(r)}{r}. \tag{3.30}$$

For a system of self-gravitating point-masses this equation has to be solved self-consistently, i.e. the density $\rho(r)$ must produce the potential $\Phi(r)$ corresponding to V_{circ}. The equation however is also applicable to a subsystem of negligible mass (i.e. the stars) that orbits in a given potential, i.e. the potential of the DM.

Integrating this over r with the additional constraint that $\rho(r) = 0$ for $r \geq r_t$ we find

$$\rho(r)\sigma^2(r) = \int_r^{r_t} \frac{\rho(r')\,V_{circ}^2(r')}{r'} dr' \tag{3.31}$$

3.4 Velocity dispersion

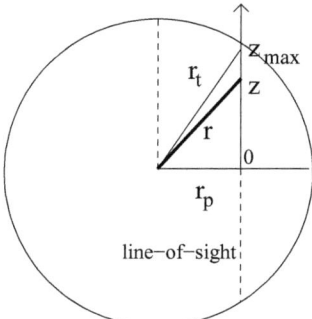

Figure 3.22: Line of sight projection at projected radius r_p through a spherical dwarf galaxy with tidal radius r_t

These 3D values have to be projected on the sky in order to be able to compare them with the observed ones.

For a spherical stellar system, - with the geometry of Figure 3.22 - we can evaluate the mean line-of-sight density

$$\rho_p(r_p) = \frac{1}{(r_t^2 - r_p^2)^{1/2}} \int_0^{(r_t^2-r_p^2)^{1/2}} dz \, \rho(r_p^2 + z^2) = \quad (3.32)$$

$$= \frac{1}{(r_t^2 - r_p^2)^{1/2}} \int_{r_p}^{r_t} dr \, \frac{\rho(r) \, r}{(r^2 - r_p^2)^{1/2}} \quad (3.33)$$

where for the last step we used the substitution $r = (r_p^2 + z^2)^{1/2}$. In the same manner we can project equation 3.30

$$\rho_p(r_p)\sigma_p^2(r_p) = \frac{1}{(r_t^2 - r_p^2)^{1/2}} \int_0^{(r_t^2-r_p^2)^{1/2}} dz \, \rho((z^2 + r_p^2)^{1/2}) \, \sigma^2((z^2 + r_p^2)^{1/2}) =$$

$$= \frac{1}{(r_t^2 - r_p^2)^{1/2}} \int_0^{(r_t^2-r_p^2)^{1/2}} dz \int_z^{r_t} dr \, \frac{\rho(r) \, V_{circ}^2(r)}{r} =$$

3. Satellite Galaxies of the Milky Way

interchanging the integrals and updating the integration boundaries gives

$$
\begin{aligned}
&= \frac{1}{(r_t^2 - r_p^2)^{1/2}} \int_{r_p}^{r_t} dr \int_0^{(r_t^2 - r_p^2)^{1/2}} dz \; \frac{\rho(r) \, V_{circ}^2(r)}{r} = \\
&= \frac{1}{(r_t^2 - r_p^2)^{1/2}} \int_{r_p}^{r_t} dr \; \frac{\rho(r) \, V_{circ}^2(r)}{r} \int_0^{(r_t^2 - r_p^2)^{1/2}} dz = \\
&= \frac{1}{(r_t^2 - r_p^2)^{1/2}} \int_{r_p}^{r_t} dr \; \frac{\rho(r) \, V_{circ}^2(r)}{r} (r_t^2 - r_p^2)^{1/2} \quad (3.34)
\end{aligned}
$$

The line-of-sight velocity dispersion σ_p is then finally

$$
\sigma_p^2(r_p) = \frac{\int_{r_p}^{r_t} dr \; \rho(r) \, V_{circ}^2(r) \, (r^2 - r_p^2)^{1/2} \, / \, r}{\int_{r_p}^{r_t} dr \; \rho(r) \, r \, / \, (r^2 - r_p^2)^{1/2}}. \quad (3.35)
$$

3.5 Comparison

3.5.1 Central velocity dispersions

With this formalism we can now calculate the expected the central velocity dispersion of the stellar distribution of the observed dwarf spheroidals (table 3.7 and equation 3.35) within the gravitational potential of the simulated subhaloes (Figure 3.15 and equation 3.13).

As it is impossible to tell *which* stellar distribution may reside in which type DM halo, we compute the central velocity dispersion values for *each* stellar distribution of the nine dwarf spheroidal galaxies orbiting in *all of* of our 30 most massive subhaloes of GA3n. We then count – for each galaxy – the number of subhaloes for which the predicted central velocity dispersion is *larger* than observed. This number is the number of subhaloes that are more massive than the real DM haloes of the dwarf galaxies.

We have carried out this exercise twice. Once, using the circular velocity profiles of equation 3.13 with the values of a obtained from our simulation, once using $a = 0.16$ for all subhaloes. This value is an upper bound for the subhalo halo profiles from Hayashi et al. (2003). The resulting profiles resemble that of the most concentrated of their subhaloes. This is conservative for our purpose.

The numbers in Table 3.7 show a remarkable result. For the GA3n subhaloes, all 11 of the satellites of the Milky Way can be accommodated within the 21 most massive subhaloes, and all but two can be accommodated within the 10 most massive.

Assuming the conservative estimate from Hayashi et al. (2003) all dwarfs can be accommodated within the 22 most massive subhaloes and all but three can be accommodated within the first 10 (all but two if the higher dispersion value is adopted for Sagittarius).

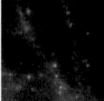

3.5 Comparison

	N_{GA3n}	$N_{Hayashi}$
Sagittarius	7(0)	15(1)
Fornax	9	18
Leo I	0	1
Sculptor	0	2
Leo II	0	0
Sextans	21	22
Carina	1	8
Ursa Minor	0	0
Draco	0	0

Table 3.7: Number of simulated subhalo profiles that predict a *larger* central velocity dispersion than that observed, i.e. the number of the 30 most massive simulated subhaloes that are *more* massive than the DM haloes of the observed dwarf spheroidals.

3. Satellite Galaxies of the Milky Way

Given that substantial realization-to-realization fluctuations are expected in the properties of the more massive subhaloes, there is a surprisingly good agreement between the kinematics of the observed satellites and those predicted by our ΛCDM simulation.

In contrast to Moore et al. (1999) and Klypin et al. (1999) we therefore find that the observed DM haloes are *not* required to be less massive than those found in our simulation.

3.5.2 Velocity dispersion profiles

For Fornax and Draco (section 3.3.2) an even closer comparison is possible. Figure 3.23 shows the predicted velocity dispersion profiles for the 20 most massive subhaloes (dotted). They are plotted over the data from Figure 3.21. The two best fitting profiles are shown in thick lines. The dashed lines show the corresponding circular velocity curves. The dot-dashed line shows the profile of the stellar distribution (values on the right axis).

The agreement is remarkably good. Not only the value of the central velocity dispersion, but also the shape of the dispersion profile can be correctly predicted. Within the given uncertainties, this is equally true for the profiles taken directly from our simulation GA3n and for the most concentrated profiles of Hayashi et al. (2003).

This result means also, that, although the velocity dispersion profile is flat, the mass distribution corresponding to equation 3.13 is in very good agreement with the observations. This mass distribution differs strongly from NFW form, and even more from an isothermal mass distribution.

The predicted peak velocities of the circular velocity curves for the models of Fornax are in the range of 33 to 51 km/s. The ones for Draco in the range of 48 to 68 km/s. These values are higher than the central velocity dispersions of Fornax and Draco by factors substantially larger than the factor of $\sqrt{3}$ that would be correct if the stars had the same radial distribution as the DM.

It is a curious that the observational data requires Fornax and Draco to have more or less similar DM haloes, i.e. rather similar peak velocity values, despite the fact that their luminosities differ by a factor of almost 60.

Very recently Klessen, Grebel & Harbeck (2003) tried to fit a non-DM dominated model to Draco using data from the Sloan Digital Sky Survey (SDSS). As no model was able to fit all the requirements they concluded – as we do – that the Draco dwarf spheroidal galaxy is strongly DM dominated.

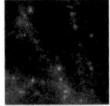

3.5 Comparison

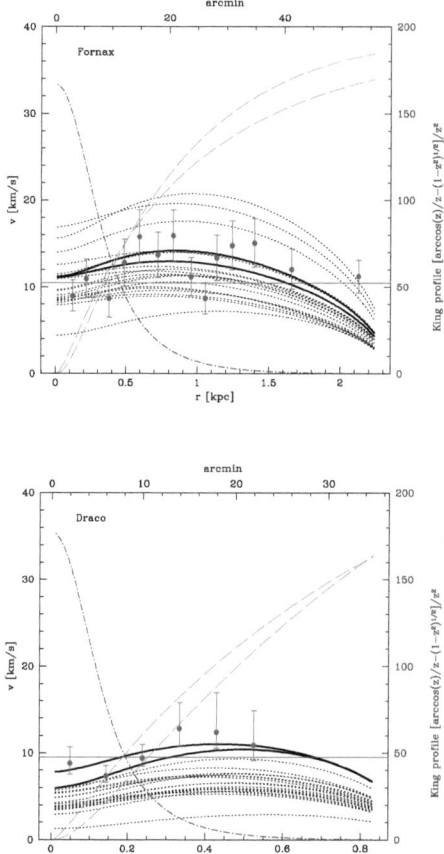

Figure 3.23: Predicted velocity dispersion profiles (dotted) for Fornax and Draco using the DM profiles of our 20 most massive subhaloes on GA3n. The two best fitting models for each dwarf spheroidal are shown with thick solid lines. The dashed lines correspond to the circular velocity curves of these best-fit dispersion profiles. King-profile fits to the stellar distributions are shown in dot-dashes (values given at the right axis). The horizontal line corresponds to the central dispersion value given by Mateo (1998). The data are from Mateo (1997) and Kleyna et al. (2001).

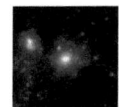

3. Satellite Galaxies of the Milky Way

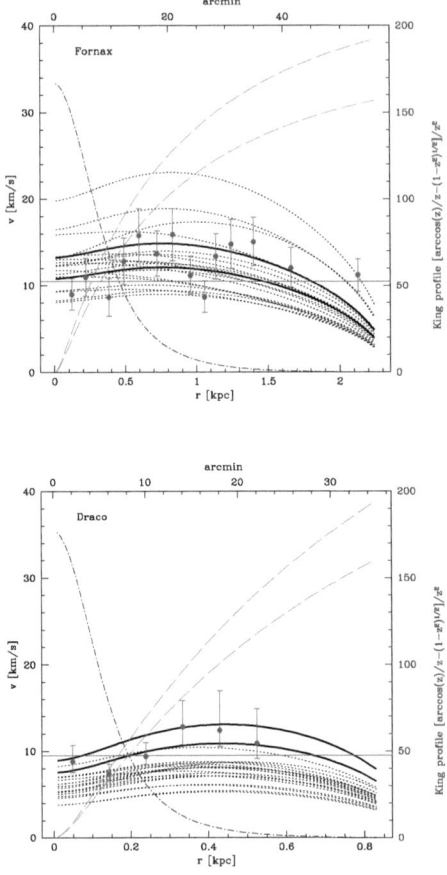

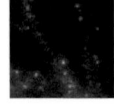

Figure 3.24: As Figure 3.23 but for the DM profiles using the conservative fits to Hayashi et al. (2003) setting $a=0.16$ for all substructure profiles.

3.6 Conclusions

The last two sections show that the potentials of the most massive subhaloes in our ΛCDM simulation are in excellent agreement with the observed kinematics of the Milky Way's satellites. There is some reason to be cautious, because our subhalo concentrations could be significantly underestimated as a result of numerical limitations, in particular the relatively large gravitational softening and particle mass in even our highest resolution simulation. We have argued from a comparison of simulations with differing resolution that the induced bias in the values of $V_{\max}$ and $r_{\max}$ is small for objects as massive as those which concern us here. Furthermore, our subhalo circular velocity curves agree well at small radii with those which Hayashi et al. (2003) obtain from simulations of the tidal stripping of individual satellites. These authors explain why tidal effects produce objects with *lower* central concentration than isolated haloes of similar maximum circular velocity. Even adopting the most concentrated profile consistent with their simulations has only a minor effect on our analysis. (Compare the rankings for the two cases in Table 3.7.) On the basis of the currently available observational and simulation data, it seems more appropriate to consider the observed kinematics of the Milky Way's satellites as a triumph for the ΛCDM model than as a crisis.

If the observed dwarf spheroidals do indeed have dark haloes of the kind that both we and Hayashi et al. (2003) suggest, there are a number of interesting consequences. In the first place, the measured "tidal radii" can have nothing to do with tidal effects, but must reflect an edge to the visible stellar population within a much more extended dark halo. This kind of structure appears to be a direct and inevitable consequence of the flat or rising velocity dispersion profiles measured in Fornax and Draco. Tidal tails and extra-tidal stars should not, then, be present in most systems.

On the other hand we know that disruption will occur inevitably in some systems. The stars of these subhaloes are tidally stripped and may be detected in streams today (Helmi & White 2000). Although there is no problem accommodating a few disrupting objects like Sagittarius and perhaps Carina (Majewski et al. 2000) it would become uncomfortable if tidal stripping were detected unambiguously in most of the other systems.

A second consequence is that the total mass associated with the dwarf spheroidals is much larger than usually assumed. The tenth most massive subhalo in our GA3 simulation has a bound mass of $1.6 \times 10^9 M_\odot$ and its progenitor system was several times more massive just before it fell into the Milky Way's halo. This sharpens the problem in understanding why the dwarf spheroidals have formed stars with such low efficiency; their stellar masses lie in the range 10^6 to $10^8 M_\odot$. Clues presumably lie in their comparatively narrow ranges of size and velocity dispersion, and in their surprisingly varied star formation histories (Mateo 1998). Finally, if the observed satellites do occupy the most

3. Satellite Galaxies of the Milky Way

massive subhaloes in a ΛCDM model, then many smaller subhaloes are presumably also present but are devoid of stars. It is interesting to look for observable consequences of their existence. Dynamical effects, for example heating or distortion of the Galactic disk or scattering of the orbits of visible halo objects, are weak, and are dominated by the most massive subhaloes rather than by their more abundant low mass brethren. Such effects may, nevertheless, be detectable in favourable cases (Font et al. 2001; Johnston et al. 2002). It may also be possible to detect substructure in gravitational lens galaxies through the statistics of flux ratios in samples of multiply imaged quasars (Mao & Schneider 1998; Chiba 2002; Dalal & Kochanek 2002; Kochanek 2003); again the effects are dominated by the few most massive subhaloes. If dark matter detection experiments are successful, it may become feasible to search for structure in the dark matter distribution at the Earth's position. Detailed analysis suggests that in a ΛCDM model the signal will be difficult, but perhaps not impossible, to see (Helmi et al. 2002). Even detecting substructure from possible DM annihilation in the near future might be possible (see next chapter).

The fact that the core structure of satellite subhaloes in a ΛCDM model agrees with the kinematics of observed satellites is *prima facie* evidence against dark matter with modified properties, for example, Warm dark matter or Self-Interacting dark matter. In recent work, the primary motivation for considering such modifications to the microscopic physics of the dark matter particles has been to reduce the concentration of haloes and the abundance of substructure (e.g. Spergel & Steinhardt (2000); Yoshida et al. (2000); Bode et al. (2001b)). A significant reduction of the central concentration of satellite subhaloes in our simulations would, however, make it difficult to produce satellites with velocity dispersions *as large* as those observed.

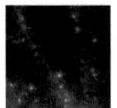

4 Dark Matter Annihilation

4. Dark Matter Annihilation

4.1 Introduction

The nature of the dark matter (DM) in the Universe is one of the most prominent unsolved questions in cosmology. As mentioned in section 1.2.2, there is evidence for non-visible matter on all scales and most of it must be composed of some as yet unknown non-baryonic particles. This follows from measurements of the actual matter density and from the predictions of Big Bang Baryon Nucleosynthesis, both of which are considered well established. Much effort is underway to find the DM particle. Although there is some chance that it can directly be detected in particle accelerators like LHC at CERN, many theoretical models propose candidates with masses out of reach of these experiments.

For this reason, other techniques for detecting DM particles in the local Universe have been pursued. One of them is to search for γ-radiation from self-annihilation of the candidate particles. As this signal scales with the square of the DM density, high density regions within the Milky Way halo, both at the centre of the Galaxy and at the centres of its satellite galaxies, are relevant.

In addition to sources in our own Galaxy, extragalactic sources, especially nearby galaxies (M87, Baltz et al. (2000) and M31 Nuss et al. (2002)) have been considered. Due to the large distance from our observing position, however, the corresponding signals are expected to be very weak. Recently, the homogenous extragalactic background of γ-annihilation radiation, produced by structure formation as a whole, has been investigated (Ullio et al. 2002).

Finally, other very strong DM concentrations, for example around the central black hole of our Galaxy, could provide a strong annihilation signal. Most recent studies suggest however, that the corresponding contribution is rather low (Merritt et al. 2002).

Depending on the nature of the annihilating particle, its annihilation flux can be estimated and its detectability with current or planned γ-ray telescopes assessed. Detectability estimates have varied a lot during the last few years. This is partly due to the fact that the matter density, previously assumed to be $\Omega_m = 1$, turned out to be lower: $\Omega_m \approx 0.3$. In addition, different assumptions about the structure of DM haloes strongly effect the total radiation prediction. Uncertainties of a factors of 1 to 100 have been discussed (Bergström et al. 1998; Baltz et al. 2000).

As the annihilation radiation signal depends on the DM density distribution, N-body simulations of DM structure formation are an ideal tool for a study of DM self-annihilation. They readily return the DM density distribution for given cosmological initial and boundary conditions. For the study of annihilation radiation in this chapter we use the high resolution simulations presented in chapter 3. This series includes the highest resolution simulation of a galaxy halo to date.

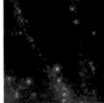

The chapter is organised as follows. In section 4.2 we give a short overview of candidate particles, supersymmetry, DM detection experiments and the annihilation process itself. We then discuss the properties of our simulations as they affect the DM annihilation (section 4.3), compute the luminosities from the smooth Milky Way halo (section 4.4) and from the substructure (sections 4.5 and 4.6). We briefly analyse the distribution of the subhaloes in our highest resolution simulation (4.7) and present the detectability computation in section 4.8 and 4.8.1. We conclude in section 4.9.

4.2 Dark matter

4.2.1 Candidate particles

The candidates for dark matter, from elementary particles to black holes, span about 75 orders of magnitude in mass. The most important candidates for baryonic DM are brown-dwarfs, "Jupiters" and 100 $M_\odot$ black-holes. The bulk of the dark matter however is probably made of non-baryonic DM particles. They split into three classes: Neutrinos, Axions and WIMPS.

Heavy tau and/or muon neutrinos (2 to 100 eV) once seemed to be good candidates, especially because they are known to exist. Neutrinos however are too light. They cannot make up all the required DM density and simultaneously fulfil the requirements of the current best-fit CDM scenario (Hut & White 1984). Axions, non-thermal relics from the Big Bang, are predicted from a newly introduced symmetry in QCD that is used to solve the CP violation problem (Peccei & Quinn 1977).

Another very well motivated DM candidate class is a weakly interacting massive particle (WIMP). The idea is that WIMPs were thermally created during the Big Bang and decoupled from the photon field at temperatures corresponding to the WIMP mass. Whereas annihilation and creation were in equilibrium at high temperatures, as the density dropped, WIMP annihilation became rare because the probability of a WIMP to find another one became small. The WIMP number density "froze out" and so a large number of WIMPs remains today.

The thermally averaged annihilation cross section $\langle \sigma v \rangle$ is roughly equal to

$$\langle \sigma v \rangle = \frac{10^{-26}\ cm^3 s^{-1}}{\Omega_{WIMP}} \quad (4.1)$$

where $\Omega_{WIMP} = \Omega_{DM}$ if the WIMP is the DM particle. The most probable WIMP candidates have masses in the range of 50 to 10^4 GeV/c^2.

4. Dark Matter Annihilation

4.2.2 Supersymmetry

In most of the theories that try to combine quantum field theory and gravity the introduction of a new symmetry is required. This "supersymmetry" is a symmetry between particles with and without spin, i.e. fermions and bosons, respectively. Postulating the existence of such a symmetry, every particle with spin must have a related "super-particle" without spin. Although this symmetry doubles the number of particles and none of the new particles has been detected so far, it is very popular. The reason is that it overcomes the "gauge hierarchy problem". It turns out that severe fine-tuning of the parameters of the standard particle model is necessary to tame large quantum corrections to the Higgs mass. The new symmetry would magically remove the necessity of this fine-tuning. Unwanted terms in the perturbation approach cancel because of the new particles that differ only in their spin. This is considered much more "natural".

Although none of the new particles yet has been detected they have been given names: The no-spin-partners of the fermions get the prefix "s" (selectron, squark, sneutrino, etc.) and the spin-partners of the bosons get the ending "-ino" (photino, gravitino, etc.). The super-particles that have the same quantum numbers can mix. The lightest of these mixing states is stable if the symmetry is conserved. This mixing state is called the neutralino. The self annihilation of this WIMP is considered in this chapter.

The straight-forward supersymmetric generalisation of the standard model of particle physics is called the Minimal Supersymmetric Standard Model (MSSM). "Minimal" means that the number of new fields (superfields) and interactions is kept as small as possible. We will use predictions of this model to compute the detectability of the WIMP-signal in section 4.8.

4.2.3 Dark matter search

If the mass of the WIMP is accessible with accelerator experiments, this technique surely is the most suited for the detection. The chances are good, especially with the new hadron collider LHC at CERN. It is supposed to reach energies of 16 TeV and will start operating in 2005. However, the energies reachable in the laboratory are limited and it may well be that the only remaining possibility for the detection of the WIMP particles is to observe those that exist in the Universe already.

There are good chances that the particles that make up the halo of the Milky Way can be detected directly by elastic scattering in suitable detectors on Earth. Due to the weak interaction, the event rates are very low (of the order of 10^{-4} events per year and per kg of detector material). In addition, the deposited energy is very small (of the order of keV).

For this reason, a growing effort has been dedicated to detecting WIMPs through the

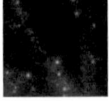

energy deposited by elastic WIMP-nucleon scattering in massive, cryogenically cooled bolometers where the phonons of a scattered WIMP can be measured. The first results of these detectors seem promising, but building large detectors still remains a huge challenge (Klapdor-Kleingrothaus & Zuber 1997).

A different detection strategy is possible if WIMPS are Majorana particles. In this case, pair-annihilations can occur, producing high-energy neutrinos, positrons, antiprotons and γ-rays. The resulting γ-ray fluxes might be detectable with current or next-generation γ-ray telescopes.

Two γ-ray signatures are produced. When the WIMPs annihilate directly into two γ-rays or into one γ-ray and a Z° Boson, very sharp line signals are produced. Although this signal would be a "smoking" gun for the presence of WIMPs, the line-flux is very small. Many more photons are produced when the WIMPs annihilate into quarks which fragment into π^0 mesons which then decay to γ-rays. This signal is much easier to detect. It lacks however a clear feature so that a careful discrimination between the possible DM annihilation and other processes producing γ-rays (cosmic rays, accelerated electrons) is required.

Both ground-based Air-Shower-Čerenkov (ACT) and space-borne telescopes might be able to detect annihilation γ-rays. When high energetic γ-rays hit the atmosphere of the Earth, they induce a shower of secondary particles which themselves decay mainly into photons, muons and neutrinos. The photons create e^+e^- pairs which cascade their energy down by BREMSSTRAHLUNG and Compton-scattering. The Čerenkov light can be detected on the ground in moonless nights with photo-multipliers. The discrimination between γ-ray induced showers and showers from cosmic-rays can be done by measuring the form of the shower on the ground. The Čerenkov light cones from γ-rays are much narrower than the cosmic ray induced ones. This discrimination is crucial for the observation, as less than 1% of the detections stem from γ-rays.

Space-borne telescopes such as EGRET and the planned GLAST satellite instrument, detect γ-rays by pair-production in the detector material. To discriminate the γ events from cosmic rays, an anti-coincidence detector sensitive to charged particles is built around the calorimeter.

Space-borne telescopes have the advantage of a rather low energy threshold, typically in the MeV range. In this low energy range many photons are expected. Typical ACT thresholds are of the order of 50 GeV. On the other hand, the effective area of space borne detectors is tiny ($< 1\text{m}^2$) compared to the large area covered by ACTs (typically 10^4m^2).

The γ-ray observations can be used to constrain WIMP parameters, in particular the self-annihilation cross-section and the particle mass. However, so far neither technique has detected a DM particle.

4. Dark Matter Annihilation

4.2.4 Annihilation

Predictions of the expected flux from the annihilation of DM particles require not only the particle parameters but also a detailed knowledge of the structure of regions of high DM density, i.e. of DM haloes. This is because the annihilation flux (in $photons/cm^2/s$) may be written as:

$$F = \frac{N_\gamma \langle \sigma v \rangle}{2\, m_{DM}^2} \int_V \frac{\rho_{DM}^2(\mathbf{x})}{4\pi\, d^2(\mathbf{x})}\, d^3 x\,, \qquad (4.2)$$

where N_γ is the number of photons produced per annihilation, $\langle \sigma v \rangle$ is the averaged product of cross-section and relative velocity, ρ_{DM} is the DM density, V is the halo volume, m_{DM} is the mass of the DM particle and d the distance from each point in the halo to the observer. The density squared weighting of the integrand in this equation results in most of the flux in dark matter haloes being produced by the small fraction of their mass in the densest regions.

As mentioned, two specific regions have been suggested as dominating the annihilation signal from haloes. A large contribution could come from the innermost part of the halo. For a distant spherically symmetric system equation 4.2 becomes

$$F = \frac{N_\gamma \langle \sigma v \rangle}{2\, d^2 m_{DM}^2} \int_0^{r_{200}} \rho_{DM}^2(r) r^2\, dr\,, \qquad (4.3)$$

so if the inner density profile is $\propto r^{-1.5}$ or steeper, the emitted flux diverges at the centre. A lower cut-off must then be specified on physical grounds, for example at the point where the annihilation timescale for the DM becomes equal to the lifetime of the halo. This divergent case may be relevant since at least some high resolution numerical simulations have suggested that the inner cusps of DM haloes could be this steep (Moore et al. 1999; Calcanéo-Roldán & Moore 2000), but contrast with Power et al. (2003) and our own simulation data in section 4.3).

A second contribution can come from small-scale structure in the DM distribution in the halo. It is now well known that 5 to 10% of the halo mass in Cold Dark Matter (CDM) haloes is contained in gravitationally self-bound substructures (Moore et al. 1999; Klypin et al. 1999). If the central regions of these subhaloes are dense enough they may produce a substantial fraction of the total annihilation radiation from the halo (Calcanéo-Roldán & Moore 2000).

In recent years, advances in integrator software, multi-mass initial condition techniques and computer speed have made it possible to simulate the DM halo of the Milky Way with sufficient resolution to see the dense structures which would dominate the annihilation signal.

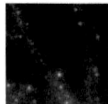

4.3 N-body simulations

In this chapter we use the N-body simulations presented in chapter 3 to predict the annihilation flux from a CDM halo similar in mass to the Milky Way's halo. Figure 4.1 shows the projected density weighted DM distribution which corresponds to an image of its annihilation radiation. The upper panel shows the distribution in linear scale, the lower panel in logarithmic scale. The total luminosity from the centre of the galaxy outperforms by far that from the subhaloes. The plotted region has a sidelength of one times the virial radius of the galaxy, i.e. 270 kpc.

The use of simulations for this approach has the advantage over semi-analytic techniques (Taylor & Silk 2003; Tasitsiomi & Olinto 2002) that the full history, dynamics and structural change of the main halo, but especially of the substructure haloes, is available. As the annihilation luminosity is strongly dependent on the structure of the highest density regions, which are by construction the ones that are the most difficult to obtain, any approximation can have large influence on the detectability prediction.

4.4 Smooth halo luminosity

The most crucial parameter determining the annihilation rate in the smooth halo is the point at which the slope of its density profile passes through the critical value -1.5. Most of the annihilation radiation will come from this region. As it is difficult to distinguish the slopes in logarithmic plot of density against radius, we use circular velocity curves. In Figure 4.2 we show circular velocity profiles

$$V_c(r) = \left(\frac{G\,M(<r)}{r}\right)^{1/2} = \left(\frac{G}{r}\int V(r)\rho_{DM}(\mathbf{x})\,dV\right)^{1/2}, \qquad (4.4)$$

where $V(r)$ is the region within distance r of halo centre.

As mentioned in section 3.2.2.1 we find the NFW profile

$$\rho(r) = \frac{3H_0^2}{8\pi G}\frac{\delta_c}{cx\,(1+cx)^2} \qquad (4.5)$$

with $x = r/r_{200}$ and $c = 10$ to be a reasonably good fit to our data outside the inner core region. A better fit is a parabolic function of the form proposed by SWTS with a width parameter of $a= 0.17$ (the dotted curve in the figure, $a=0.074$ for equation 3.13). This profile has a substantially shallower density profile at small radii than the NFW profile. The measured circular velocity profiles of GA2n and GA3n agree very well, but a comparison with GA1n and GA0n suggests that this apparent convergence is in part a fluke. The curves for the two lower resolution simulations converge to within 5%

4. Dark Matter Annihilation

Figure 4.1: *Upper panel:* The distribution of DM in our highest resolution simulation GA3n. The region displayed is a cube of side 1080 kpc, i.e. four times r_{200}. Each particle is weighted by its local density so that the picture represents an image in linear annihilation radiation. *Lower panel:* The same image in logarithmic scale.

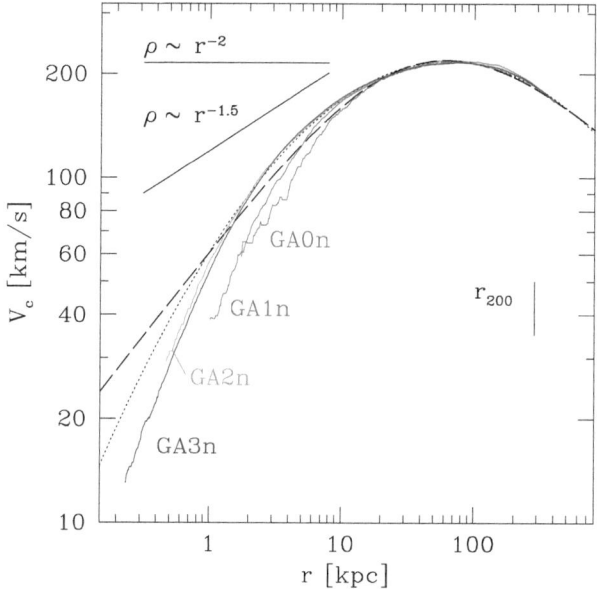

Figure 4.2: Circular velocity curves for the simulations GA0n, GA1n, GA2n and GA3n. The vertical line indicates the location of the virial radius r_{200}. The best fit NFW profile with concentration $c = 10$ is plotted in long dashes. A fit of the form proposed by SWTS with $a=0.17$ is shown in dots ($a=0.074$ for equation 3.13). At small radii, the slope for GA3n is considerably below that corresponding to a density profile with $\rho \propto r^{-1.5}$.

4. Dark Matter Annihilation

of the high resolution answer beyond about 5 times their respective softening lengths. Assuming this to be true for GA2n and GA3n also, the inner slope of the density profile of our simulated halo is already established to be well below -1.5. With this criterion, convergence for GA3n is achieved just outside 1 kpc or at about 0.4% of r_{200}. This agrees with the expectation from the convergence study of Power et al. (2003) who performed a large number of simulations of several haloes using two different N-body codes as well as a wide variety of code parameters (timestep, softening, particle number, etc.).

The concentration parameter for our NFW fit to our halo is $c = 10$. Thus, $\delta_c = 4.48 \times 10^4$ and the above value of r_{200} implies a scale radius $r_s = 27$ kpc, and a density at the Sun's position ($r_0=8.0$ kpc) of $\rho_0 = 1.2 \times 10^7$ $M_\odot/\text{kpc}^3 = 0.46$ GeV c^{-2}cm^{-3}.

For our NFW fit, half of the annihilation radiation is predicted to come from within $0.26 r_s$ which is 7 kpc. Thus the resolution in GA3n appears easily sufficient to measure the bulk of the emission, even though still better resolution would clearly be desirable. Unfortunately, the numerical situation will not improve dramatically in the next few years unless revolutionary new techniques are discovered. As discussed at length by Power et al. (2003), an increase in halo particle number by (say) two orders of magnitude would provide an increase in length resolution at halo centre by at best a factor of ten.

Many authors have tried to determine the inner slope of dark halo density profiles through physically based analytic arguments (Peebles 1980; Hernquist 1990; Syer & White 1998; Nusser & Sheth 1999; Subramanian et al. 2000; Dekel et al. 2002) but despite some interesting ideas a convincing final answer is still missing. For the purposes of this work the critical issue is whether the slope of the density profile interior to the points for which it has so far been estimated accurately from simulations (i.e. at radii below 1 kpc) remains significantly shallower than -1.5. If it does, then the integral over the smooth halo density distribution is convergent and can be estimated reasonably accurately from high resolution simulations, for example from GA3n which is currently the best resolved simulation of a galaxy halo ever carried out.

Bergström et al. (1998) show that for the case where a smooth NFW halo, similar to that of Figure 4.2, is a good description of the Milky Way's halo, the γ-ray flux may just be detectable for some MSSM models with the next generation telescopes. This flux could, however, be significantly enhanced if the density distribution within the halo is sufficiently clumpy (Bergström et al. 1999; Calcanéo-Roldán & Moore 2000). Ullio et al. (2002) find that without clumpiness enhancing the total flux by a factor of 10 or more, the high-galactic latitude γ-ray background as measured by EGRET can not be the result of DM annihilations. We now estimate whether the clumpiness of our simulated haloes is enough to produce a large enhancement.

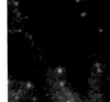

4.5 Halo substructure

In GA0n, GA1n, GA2n and GA3n the total mass in gravitationally bound substructures, identified with SUBFIND, is 1.7%, 3.0%, 5.1% and 4.1%, respectively; the fluctuations are due to the exclusion or inclusion of one or two massive satellites within the radius $r < r_{200}$ we consider to define the halo boundary. These percentages are very similar to those found in the cluster simulations of Springel et al. (2001a).

4.5.1 Subhalomassfunction

In Figure 4.3 we show the abundance of substructures as a function of mass fraction for our highest resolution simulation GA3n, as well as for S4, the highest resolution simulation of Springel et al. (2001a).

These two mass functions are remarkably similar and are very close to a power-law $dn/dm \propto m^{-1.78}$ as shown by the solid line in the figure. They are consistent with the findings of other authors (Moore et al. 1999; Klypin et al. 1999; Metcalf & Madau 2001; Font et al. 2001; Helmi et al. 2002) although the range of values (1.75 to 1.9) suggests that the close agreement between these two particular haloes is likely to be a fluke. Some variation is undoubtedly due to the fact that different authors use different algorithms to define substructure, but these effects have not yet been studied in detail. Note that such slopes are also found at low mass for the mass function of isolated haloes in a ΛCDM Universe, suggesting that the fraction of mass lost depends only weakly on the initial mass of an accreted DM halo. These slopes are shallow enough to ensure that most of the mass in substructures is contained in the few most massive objects. Thus we do not expect the total mass in substructure to increase significantly as resolution is extended below our current limit. In GA3n there are 1973 subhaloes with more than 20 particles.

As suggested by Bergström et al. (1999) and Calcanéo-Roldán & Moore (2000), if these subhaloes are sufficiently concentrated, they can make a substantial contribution to the total annihilation flux from the halo. Just as for the main halo, the critical question is the structure in their inner regions, in particular, whether they contain more or less mass at the highest densities than does the core of the main halo. Recently Hayashi et al. (2003) carried out high resolution simulations of the tidal stripping of satellites to assess how their internal structure is affected by the removal of the outer material. Their results show clearly that the stripping process reduces the density of an accreted object at *all* radii, not just in its outer regions. Thus tidal effects progressively lower the annihilation luminosity of an accreted system. If its γ-ray flux was convergent in the inner regions while it was an independent system, then it converges even more rapidly once it has become a partially stripped "satellite". Both the individual satellite simulations of

4. Dark Matter Annihilation

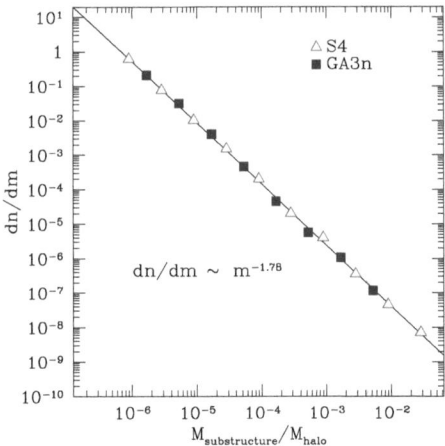

Figure 4.3: Subhalo mass function for the GA3n (Milky Way) and S4 (Cluster) simulations. Subhaloes identified by SUBFIND and with ten or more particles are included in these distributions.

Hayashi et al. (2003) and the results plotted in SWTS, suggest that the inner structure of halo substructures corresponds to density profiles shallower than NFW.

4.6 Substructure luminosity

We now proceed estimate the annihilation luminosity of our simulations directly in order to evaluate how much of the flux is contributed by the various different parts of the system. To do this we evaluate the "astrophysical" part of equation 4.2 in the form

$$J = \int_V \rho_{DM}^2 \, dV = \sum_{i=1}^{N_{200}} \rho_i \, m_i, \qquad (4.6)$$

where ρ_i is an estimate of the DM density at the position of the ith particle, m_i is its mass, and N_{200} is the total number of particles within r_{200}. With this representation of the flux, it is easy to evaluate the contribution from any sub-element of the halo simply by restricting the sum to the relevant particles. We will consider how these Js should be converted into γ-ray detectability limits for various WIMP parameters in section 4.8.

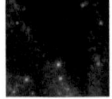

4.6 Substructure luminosity

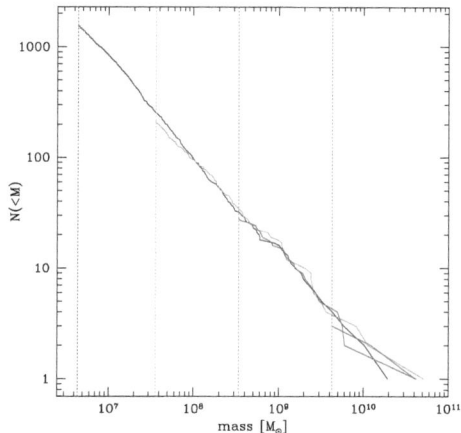

Figure 4.4: Cumulative number of subhaloes in the four simulations GA0n, GA1n, GA2n and GA3n down to 20 particle subhaloes.

4.6.1 Density estimation

The quality of our estimate of J will obviously depend strongly on the quality of our estimates of DM density. For a given simulation we would like these estimates to have the maximum possible resolution. We have elected to determine the ρ_i by Voronoi-tessellation. This procedure uniquely divides three dimensional space into convex polyhedral cells, one centred on each particle, defined to contain all points closer to that particle than to any other. Figure 4.5 shows a two-dimensional example.

The density estimate for each particle is then its mass divided by the volume of its cell. We have used the publicly available package QHULL[1] to make these estimates. One major advantage of this scheme in comparison, say, to density estimation with an SPH kernel is that it is parameter-free and has very high resolution; the density estimate for each particle is determined by the position of its few nearest neighbours. Another is that it is unbiased and that the sum $\sum_i^{N_{200}} m_i/\rho_i$ recovers the full volume.

---

[1] www.geom.umn.edu/software/qhull/

4. Dark Matter Annihilation

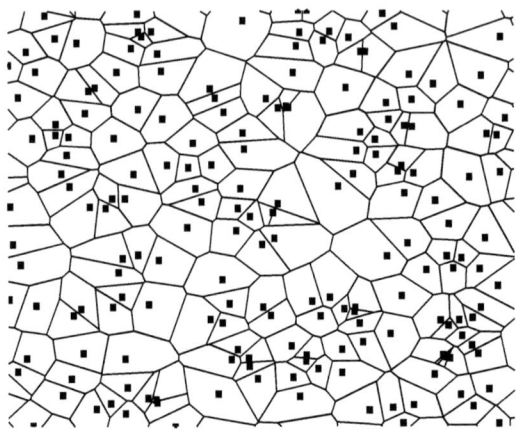

Figure 4.5: 2-dimensional example of a space-decomposition with Voronoi-tessellation. The particles are shown by squares.

4.6.2 Enhancement due to substructure

4.6.2.1 Flattening and discreteness

A direct numerical evaluation of J using equation 4.6 will differ from the value obtained by carrying out the appropriate integral over the circular velocity curves of Figure 4.2. This is because any deviation from strict spherical symmetry results in the sum of $m_i \rho_i$ over a spherical shell being larger than the product of its total mass and its mean density. Thus flux "enhancements" will result from bound subhaloes, from unbound streams, from the overall flattening of the halo and from noise in the density estimates due to discreteness effects.

The enhancements due to flattening and discreteness effects can be estimated from DM haloes with random particle positions set up to have an NFW density profile. Such haloes can be constructed as follows.

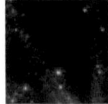

4.6 Substructure luminosity

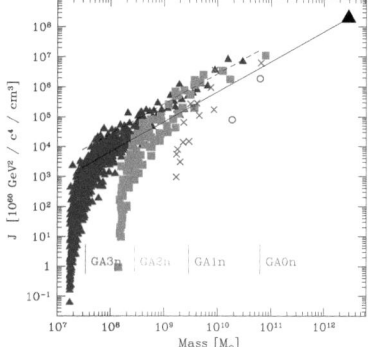

Figure 4.6: The sum J, which is proportional to the expected annihilation luminosity, is plotted as a function of subhalo mass for all subhaloes with more than 20 particles in GA0n (circles), GA1n (crosses), GA2n (dots) and GA3n (triangles). The values of J for the main haloes as a whole are also indicated using the larger symbols. Subhaloes with the same J/M as the GA3n main halo would lie on the solid line. The dashed line corresponds to a 4 times larger value of J/M. Vertical lines indicate subhalo masses corresponding to 100 particles in each of the four simulations. Above these limits subhaloes in all four simulations have similar values of J/M with no obvious trend with subhalo mass.

In spherical coordinates and for a spherically symmetric density distribution

$$\rho = \frac{dM}{dV} = \frac{m\,dn}{dV} \qquad (4.7)$$

the number density is

$$dn = \frac{1}{m} d\varphi \, \sin(\theta) \, d\theta \, \rho(r) r^2 dr \qquad (4.8)$$

We draw linearly distributed random numbers x, y and z in the interval $[0..1]$. They have to be mapped on the variables φ, θ and r. This can be done by integrating the differential mapping and solving for the original variables. For φ this is trivial. For θ we find

$$dy = \sin(\theta) \, d\theta \;\Rightarrow\; y = -\cos(\theta) \;\Rightarrow\; \theta = \arccos(-y) \qquad (4.9)$$

4. Dark Matter Annihilation

For the interval $[0..1[$ for the random numbers, the mapping is

$$\begin{aligned} \varphi &= 2\pi \; [0..1[\\ \theta &= \arccos(-1 + 2\;[0..1[) \end{aligned} \quad (4.10)$$

For constant density the z-mapping is reversible

$$dz = r^2 dr \;\Rightarrow\; z = \frac{1}{3}r^3 \;\Rightarrow\; r = \sqrt[3]{3\,z}$$
$$r = r_{vir}\sqrt[3]{[0..1[}. \quad (4.11)$$

For a NFW halo

$$z = \left[\frac{\ln\left[1 + \frac{c}{r_{max}}r\right] + 1/(1 + \frac{c}{r_{max}}r) - 1}{\ln[1 + c] + 1/(1 + c) - 1} \right] \quad (4.12)$$

which can not be reversed but allows one to produce a look-up table: z runs from $[0..1]$ for r from 0 to r_{max}. With equations 4.10 and 4.12, the particle positions $(\sin(\theta)\cos(\varphi)\,r, \sin(\theta)\sin(\varphi)\,r, \cos(\theta)\,r)$ can be obtained for the desired number of particles.

For our artificial NFW haloes we find that in any region where the mean particle separation is small compared to the length-scale for density variations, our Voronoi scheme will give 22% more flux than what is estimated from average densities in spherical shells. For an ellipsoidal NFW halo with axial ratios 1:1.2:1.8, similar to what we measure in the inner regions of our simulated haloes, the enhancement due to the flattening is about 15%. The enhancement due to bound structures is estimated explicitly below.

4.6.2.2 Selfsimilarity

If subhaloes were simply scaled down copies of the main halo, then their fractional contribution to the annihilation luminosity of the system would be the same as their fractional contribution to its mass, i.e. roughly 5%. However, the algorithm SUBFIND bounds substructures at the point where their density equals the local density of material within the main halo.

As a result, the internal structure of a subhalo cannot be similar to the main halo as a whole, but might be similar to that part of it which lies interior to the subhalo's position (scaled down in size by the cube root of the ratio of the substructure's mass to the total halo mass within a sphere passing through it). If such self-similarity were actually to hold then the annihilation luminosity per unit mass (i.e. the quantity J/M) would be the same for the subhalo as for the main halo interior to its position. In fact, however, the study of Hayashi et al. (2003) shows that the density of a satellite at radii approaching its tidal limit is is reduced by a substantially greater factor than that in the regions near

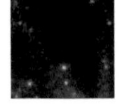

4.6 Substructure luminosity

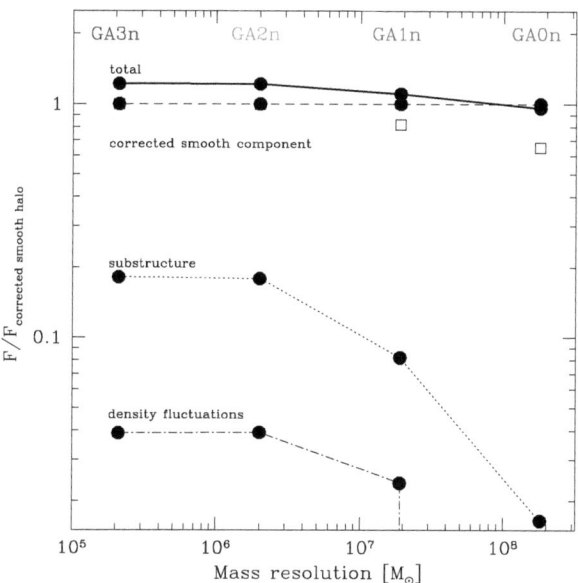

Figure 4.7: The luminosities of different halo components as a function of mass resolution. The values for each simulation are scaled so that the luminosity of its smooth halo is unity after correction for Poisson discreteness and flattening. The large squares show the luminosities of these smooth halo components in units of the value found in GA3n.

4. Dark Matter Annihilation

the peak of its circular velocity curve. This effect increases the luminosity per unit mass of a substructure relative to the main halo interior to its position. Higher values might also be expected because subhaloes had lower mass progenitors at high redshift than did the main halo, and so began life with higher concentration (see Navarro, Frenk & White (1997) and Bullock et al. (2001) for estimates). On the other hand, we have argued above that tidal effects also reduce the concentration of the inner core of subhaloes, thus reducing their annihilation luminosities. The J/M values for substructures reflect the combination of all these effects.

Figure 4.6 shows the sums J – which are proportional to annihilation luminosity – as a function of subhalo mass for all four of our simulations and for subhaloes containing at least 50 particles. A point is also plotted for the main halo as a whole. If subhaloes were similar to the GA3n main halo, they would lie along the solid line of constant J/M. Clearly, J/M is larger for the subhaloes. The vertical lines in the plot indicate subhalo masses corresponding to 200 particles for each of the four simulations.

Such GA3n subhaloes have J/M values typically 4 times larger than the main halo as a whole and twice as large as the part of the main halo interior to their position. The other effects discussed above presumably account for the remaining factor. Note that there is no indication that J/M depends on subhalo mass for subhaloes with more than 200 particles. This implies that the total luminosity from subhaloes, like the total subhalo mass, is dominated by the largest objects.

4.6.2.3 Luminosity contributions

The upper panel of Figure 4.7 shows the contributions of different components to the total estimated annihilation luminosity of our simulated haloes. The smooth halo component is shown by a dashed line. This was calculated from the circular velocity curves of Figure 4.2 and was corrected up by 15% to account for the fact that the main halo is ellipsoidal rather than spherical, and by 22% to account for discreteness noise in our Voronoi density estimates. The total luminosity from bound subhaloes is indicated by a dotted line. The remainder of the total halo luminosity is then assumed to come from unbound substructure and is indicated by a dot-dashed line. All values are given in units of the corresponding smooth halo luminosities.

The values of the latter, relative to GA3n, are indicated by boxes in the Figure. The close agreement of the circular velocity curves for GA3n and GA2n results in identical predictions for the smooth halo luminosity. The values predicted for GA1n and GA0n are only smaller by 20% and 35% respectively, suggesting that convergence is approximately achieved in the inner regions even for relatively low mass resolution. The reason is simply that the simulations predict the half-light radii of haloes to be relatively large (8.6 kpc

in GA3n) in comparison with the nominal resolution.

In GA3n, the total luminosity is a factor of 1.7 – the 'clumpiness factor' – times the value for a smooth spherical halo with the same circular velocity curve. Of this 70% increase, 25% is due to bound substructures with 10 or more particles, 22% is due to Poisson discreteness noise, 15% to the mean flattening of equidensity surfaces in the inner halo and 8% to unbound fluctuations. Although the fraction of the annihilation luminosity contributed by substructure nearly does not change between GA2n and GA3n, it is clear that this does not indicate absolute convergence.

The total mass in substructures in GA2n is 1.2 times that in GA3n (a consequence of the chance inclusion of a couple of substructures in GA2n which lie just outside r_{200} in GA3n). The luminosity per unit mass of substructures is a factor of 1.11 larger for GA3n than for GA2n. Similarly, the luminosity of the inner 10% of r_{200} (which dominate the luminosity of the smooth halo) is a factor of 1.05 larger in GA3n than in GA2n.

4.6.3 Results

In order to get an idea of an upper limit of the additional luminosity which might be found at higher resolution it is instructive to extrapolate the variation between GA1n and GA2n down to a mass resolution of one solar mass. (Note that GA3n lies well *below* this extrapolation.) Even at such high resolution, the total luminosity is predicted to be only about 3.0 times that of the smooth halo in GA3n.

Lake (1990) and Bergström et al. (1999) suggested that if dark matter substructure happens to be close to the observer, it might be easier detectable than the galactic centre itself.

This possibility was judged plausible by Tasitsiomi & Olinto (2002; hereafter TO) who assumed subhaloes to be distributed through the Galactic halo like the DM itself and tried various models for their internal structure. For the internal structure predicted by our simulations, however, it is very unlikely that any substructure will outshine the Galactic Centre. The most massive and most luminous substructures are rare and tend to avoid the inner Galaxy. They presumably correspond to the known satellites of the Milky Way (see SWTS), the nearest of which is Sagittarius, 24 kpc from the Sun. The greater abundance predicted for less massive substructures is insufficient to compensate for their lower predicted luminosities – the chance that the received flux is dominated by an unexpectedly nearby low-mass substructure is predicted to be very low.

4. Dark Matter Annihilation

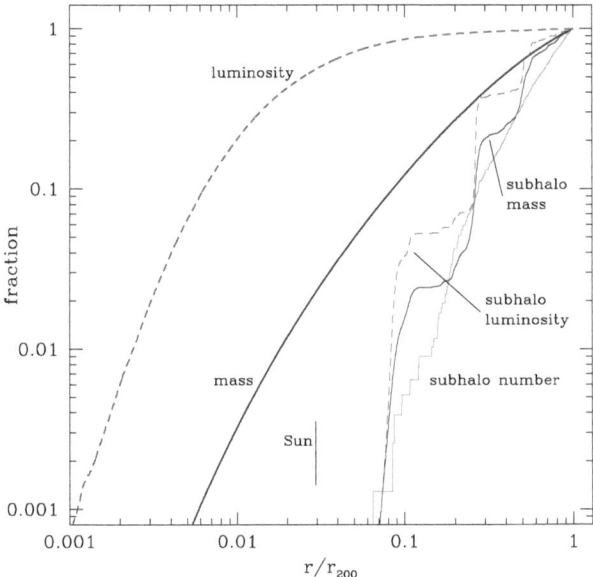

Figure 4.8: Cumulative luminosity (dashed) and mass (solid) for the GA3n main halo (thick) and for the subhaloes with more than 50 particles (thin) as a function of radius.

4.7 Subhalodistribution

These issues are clarified in figure 4.8, where we plot cumulative radial distributions for enclosed total mass and total annihilation luminosity *exclusive of substructure*, as well as for substructure number, mass and annihilation luminosity. While the diffuse luminosity is much more strongly concentrated towards the Galactic Centre than the mass, all properties of the substructure are more *weakly* concentrated. In addition, since J/M is independent of subhalo mass, which in turn is almost independent of distance from the Galactic Centre, the distributions of substructure number, mass and luminosity are all rather similar. The latter two are much noisier than the first because most of the mass and most of the luminosity come from the few most massive subhaloes. At 8 kpc, the Sun's galactocentric radius lies in the region where most of the diffuse annihilation radiation originates, but well inside any of the resolved subhaloes in GA3n (the first is at $R = 17.2$ kpc) and even further inside any of the more massive subhaloes (the first is at $R = 70$ kpc).

While Figure 4.8 was constructed directly from GA3n based on our Voronoi estimate of J/M for each particle, we obtain almost identical results if we instead use our SWTS model fits to main halo and subhalo circular velocity curves and assume that J/M is 4 times the value for the main halo for all subhaloes which are too small for circular velocity curves to be estimated. This again suggests that GA3n has high enough resolution to get reliable results for the problem at hand.

4.8 Detectability

The 148 most massive subhaloes, having masses between $5.2 \times 10^7 M_\odot$ and $1.8 \times 10^{10} M_\odot$, are well resolved enough to obtain r_{max} and v_{max}, the distance and the circular velocity at the peak of the rotation curve.

The substructure profiles are shallower than the NFW form. For an estimate we use the profile of SWTS with a fitting parameter a for each subhalo. This rotation velocity profile leads to a density profile that reaches a maximum value and has *decreasing* density if the distance to the substructure centre gets smaller than a certain small value. We use the maximum value of the density for this innermost region.

The SWTS density profile can be obtained from

$$\rho(r) = \left.\frac{dM}{dV}\right|_r = \left.\frac{d(V(r')^2 r'/G)}{d(4\pi/3\ r'^3)}\right|_r \qquad (4.13)$$

4. Dark Matter Annihilation

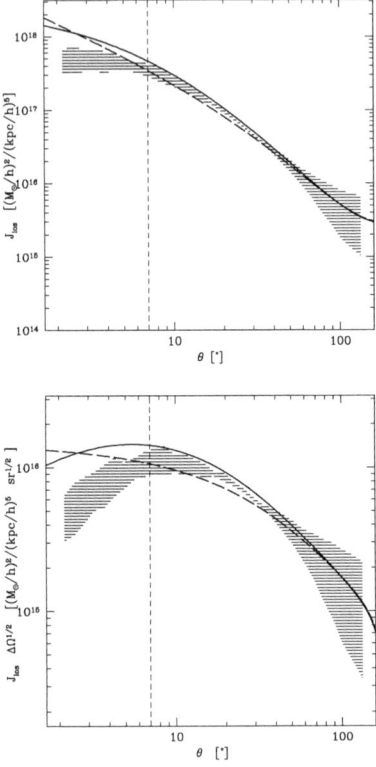

Figure 4.9: Mean predicted surface brightness of annihilation radiation as a function of angular distance from the Galactic Centre. The hatched regions enclose profiles estimated directly from six different "solar" positions 8kpc from the centre of GA3n, while the two curves gives results based on SWTS (solid) and NFW (dashed) fits to the circular velocity curve, together with an enhancement factor of 1.7. In the left-hand panel the surface brightness profiles are plotted directly, whereas on the right they are multiplied by one power of the angle θ to represent the signal-to-noise expected in an observation of fixed duration of a region whose size increases in proportion to θ, (specifically $[\theta, 1.01\theta]$). Vertical dashed lines indicate the angle subtended by the gravitational softening length at the distance of the Galactic Centre

4.8 Detectability

with $V(r')$ being the SWTS rotation curve of equation 3.13. It may be written as

$$\rho_{SWTS}(r) = \frac{V_{max}^2}{4\pi G} e^{-2a\ln(\frac{r}{r_{max}})^2} \frac{1}{r^2} \left[1 - 4\,a\,\ln\left(\frac{r}{r_{max}}\right)\right] \quad (4.14)$$

for $r > r_{core}$ where r_{core} is the radius at which the density profile becomes flat. This radius can be obtained by requiring that the mass interior to r_{core} (at constant density) is the mass required to produce the rotation curve at r_{core}, i.e.

$$\frac{4\pi}{3} r_{core}^3 \rho_{SWTS}(r_{core}) = M_{core} = \frac{V(r_{core})^2\, r_{core}}{G}. \quad (4.15)$$

We find

$$r_{core} = r_{max}\, e^{-\frac{1}{2a}}. \quad (4.16)$$

The maximal density of the profile is then

$$\rho_{SWTS}(r_{core}) = \frac{3}{4\pi G} \frac{V_{max}^2}{r_{max}^2} e^{1/(2a)}. \quad (4.17)$$

This density is adopted for $r < r_{core}$. It is easy to crosscheck, that the core radius is the radius at which the rotation curve has a slope of 1. The corresponding mass $M(<r)$ is just

$$M(<r) = \frac{V^2\, r}{G} = \frac{V_{max}^2\, e^{-2a\ln(\frac{r}{r_{max}})}\, r}{G} \quad (4.18)$$

We note however, that the density profile is not smooth at r_{core}. The radius at which the density profile gets flat is a bit smaller than r_{core} for typical values of a:

$$r_{flat} = r_{max}\, e^{-\frac{1}{8a}[1+\sqrt{9+16a}]}. \quad (4.19)$$

The density at this point is

$$\rho_{SWTS}(r_{flat}) = \frac{1}{8\pi G} \frac{V_{max}^2}{r_{max}^2} e^{\frac{3\sqrt{9+16a}-1-8a}{16a}} \left[3 + \sqrt{9 + 16a}\right] \quad (4.20)$$

Using r_{flat} instead of r_{core}, and thus obtaining a smooth density profile resulting in a sooth total rotation curve is not a bad approximation. For the GA3n halo, the relative difference in mass integrated up to the maximum of the rotation curve is 4.0×10^{-5}.

The values for the main halo of GA3n with $a = 0.0738$, $V_{max} = 228$ km/s and $r_{max} = 58.5$ kpc are $r_{core} = 0.067$ kpc and $\rho(r_{core}) = 7.36 \times 10^8$ M$_\odot$/kpc^3 = 28.0 GeV/c^2/cm^3 as well as $r_{flat} = 0.048$ kpc and $\rho(r_{flat}) = 7.48 \times 10^8$ M$_\odot$/kpc^3 =28.4 GeV/c^2/cm^3. The density at the maximum of the rotation curve is

$$\rho_{SWTS}(r_{max}) = \frac{1}{4\pi G} \frac{V_{max}^2}{r_{max}^2} \quad (4.21)$$

4. Dark Matter Annihilation

which has the value of 2.8 ×10^8 M$_\odot$/kpc^3 = 0.011 GeV/c^2/cm^3 for GA3n.

We compute the astrophysical part of the annihilation luminosity numerically from equations 4.14, 4.19 and 4.20.

For our highest resolution simulation GA3n we can make artificial sky maps of the annihilation radiation by choosing appropriate positions for the Sun within the model. Figure 4.9 shows the result of this exercise based on six possible solar positions. Even though we average the predicted surface brightness around circles of fixed Galactocentric angle, there is significant variation among the resulting profiles. This is primarily a consequence of the prolate structure of the inner regions, clearly visible in Figure 4.2. The profiles flatten out within about 10 degrees, quite possibly as a consequence of poor numerical resolution. *Prima facie* this seems plausible since the angular scale corresponding to our softening length (the vertical lines in the plots) is only 4 or 5 times smaller than the radius where the profiles bend. Some indication of the strength of this effect is given by the two curves. These indicate predictions based on SWTS (solid) and NFW (dashed) fits to the circular velocity profile of GA3n, corrected up by a clumpiness factor of 1.7. Both inward extrapolations predict substantially more flux within a few degrees than the direct numerical estimates.

It is important to note, however, that the area potentially available for a measurement at distance θ from the Galactic Centre increases as θ (for $\theta \ll \pi/2$). As a result, the counts available to detect a signal vary as θ^2 times the profiles shown in the left panel of Figure 4.9 while, for a uniform background, the noise against which the signal must be detected grows only as θ. Thus the potential S/N for a detection, shown in the right-hand panel, is given by θ times the profile. This function is quite flat out to 20 degrees, both for the directly measured profiles and for our two alternative fitting formulae. This has two consequences: (i) since for many observations the background is higher in the direction of the Galactic Centre, it may be advantageous to observe at large θ if one has a detector with sufficient field-of-view; (ii) the estimates of detectability which we give below for detectors with a wide field of view are not greatly affected by the resolution of our simulation.

4.8.1 Cross-section computation

Using the above results, we can check if the centre of the Milky Way or a substructure halo close to the Sun might be detectable with next generation γ-ray telescopes. We use the excellent package DARKSUSY[2] to compute the cross-sections $\langle \sigma v \rangle$ and neutralino masses m_χ for a Monte Carlo sampling of the MSSM parameter space. The results are shown in Figure 4.11. From roughly two million models randomly picked out of the

---

[2]www.physto.se/~edjso/darksusy/

4.8 Detectability

Table 4.1: MSSM parameter bounds and our adopted sampling strategy. (for a detailed description see Bergström et al. (1998)).

sampling	tan(β) log	μ [GeV] log	M_2 [GeV] log	m_A [GeV] log	m_0 [GeV] log	A_b/m_0 lin	A_t/m_0 lin
min	3.0	-50000	-50000	0	100	-3	-3
max	60.0	50000	50000	10000	30000	3	3

parameter space, 19421 did not violate current accelerator bounds. Of these, 825 result in relic densities of cosmological interest, i.e. $0.17 < \Omega_{DM} < 0.43$, the 95% confidence interval quoted by Spergel et al. (2003). For the MSSM parameter space we followed the choices of TO (Table 4.1).

Instead of the total flux from the source (equation 4.3) the line-of-sight integral of the DM distribution averaged over the angular resolution of the telescope has to be used.

$$F_{astro} = \frac{1}{\Delta\Omega} \int_{\Delta\Omega} \int_0^\infty dl \; \rho^2(l) \qquad (4.22)$$

The "angular resolution" σ_θ is half the opening angle of the observing cone and thus

$$\Delta\Omega = 2\pi \left(1 - \cos(\sigma_\theta)\right). \qquad (4.23)$$

Averaged over a gaussian beam of width $\sigma_\theta=0.1°$, we find that the line-of-sight integral of the square of the mass density in the direction of the Galactic Centre takes values 5.2×10^{25} and 1.8×10^{24} GeV^2/c^4/cm^5 for inward extrapolations of our NFW and SWTS fits to the main halo circular velocity curve. The large difference reflects the fact that this estimate is sensitive to density values far inside the region resolved by our simulations. Figure 4.2 suggests that the lower value obtained from the SWTS extrapolation is more likely to be correct. To estimate the maximum plausible brightness for a subhalo, we chose six possible positions for the Sun, each 8 kpc from the centre. From each position we made an artificial sky and identified the brightest subhalo as seen with a telescope with a 0.1° beam. The beam-averaged line-of-sight integral of density squared for the (apparently) brightest substructure in these six realisations is 4.9×10^{23} GeV^2/c^4/cm^5.

4. Dark Matter Annihilation

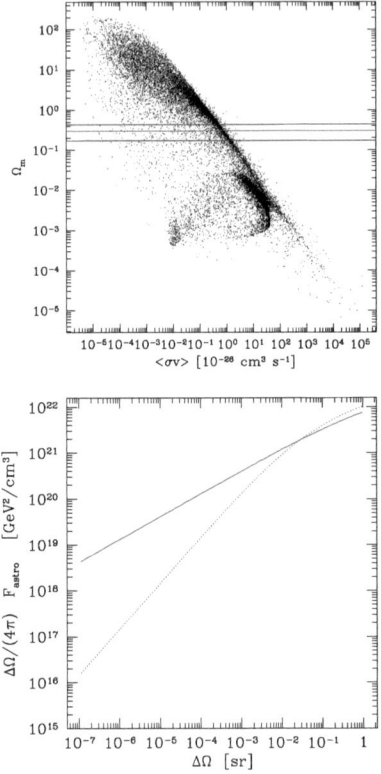

Figure 4.10: *Upper Panel:* Relic density for particle models randomly drawn from the MSSM parameter space as a function of the corresponding cross-sections. The horizontal lines show the 95% confidence interval of the WMAP results (Spergel et al. 2003). *Lower Panel:* The astrophysical part of the annihilation flux for a line of sight to the centre of the (perfectly smooth) simulated MW as a function of the angular resolution (solid angle) of the telescope. The solid line corresponds to our best NFW fit, the dotted line to our best SWTS fit. For the angular resolution of typical γ-ray telescopes (around 10^{-5} sr) the SWTS profile provides much less flux than the NFW fit in a single resolution element.

4.8 Detectability

The number of photons arriving on the telescope is

$$N_{annihilation} = A_{eff}\, t \frac{N_{cont}}{2} \frac{\langle \sigma v \rangle}{m_\chi^2} \frac{\Delta \Omega}{4\pi} F_{astro} \qquad (4.24)$$

with the telescope parameters A_{eff} (the effective area) and t (the integration time).

We concentrate on the γ-ray continuum signal which is easier to detect than the line signal (Baltz et al. 2000,TO). We use the approximation of the number of continuum photons produced in one annihilation from TO

$$N_{cont}(E_\gamma > E_{th}) = \frac{5}{6}\left(\frac{E_\gamma}{m_\chi}\right)^{3/2} - \frac{10}{3}\frac{E_\gamma}{m_\chi} + 5\left(\frac{E_\gamma}{m_\chi}\right)^{1/2} + \frac{5}{6}\left(\frac{E_\gamma}{m_\chi}\right)^{-1/2} - \frac{10}{3} \qquad (4.25)$$

where E_{th} is the low-energy threshold of the telescope and m_{chi} the mass of the neutralino. The significance of a detection is given by the number of detected photons from DM annihilations over the background

$$M_s \leq \frac{N_{annihilation}}{\sqrt{N_{background}}} \qquad (4.26)$$

For given telescope and observation parameters i.e. the effective area A_{eff}, the integration time t, the angular resolution σ_θ or the beam half-angle θ_{max}, the effective γ-ray background and the significance required for detection M_s, the smallest detectable cross-section $\langle \sigma v \rangle$ can be computed as a function of the neutralino mass m_χ (Bergström et al. 1998; Baltz et al. 2000, TO).

$$\langle \sigma v \rangle_{min} = \frac{M_s m_\chi^2 \sqrt{N_{background}}}{N_{cont}\, A_{eff}\, t\, \frac{\Delta \Omega}{4\pi}\, F_{astro}} \qquad (4.27)$$

4.8.2 Background

Bergström et al. (1998) and Baltz et al. (2000) give the following fits for the background contributions. Hadrons can produce particle showers in the earth's atmosphere which might be misinterpreted as γ-ray detections in ATC detectors. To be conservative for our purpose we completely neglect this background contribution assuming a perfect rejection mechanism.

$$\frac{dN_{hadronic}(E_\gamma > E_{th})}{dt\, dA} = 6.1 \times 10^{-3} \left(\frac{\Delta \Omega}{1 sr}\right) \left(\frac{E_{th}}{1\text{ GeV}/c^2}\right)^{-1.7} \text{cm}^{-2}\,\text{s}^{-1} \qquad (4.28)$$

For ACT observations we however do include electron induced showers:

$$\frac{dN_{electron-induced}(E_\gamma > E_{th})}{dt\, dA} = 3.0 \times 10^{-2} \left(\frac{\Delta \Omega}{1 sr}\right) \left(\frac{E_{th}}{1\text{ GeV}/c^2}\right)^{-2.3} \text{cm}^{-2}\,\text{s}^{-1} \qquad (4.29)$$

4. Dark Matter Annihilation

All observations at low latitudes are affected by the diffuse Galactic γ-ray background observed by EGRET. Its origin is probably the interaction of cosmic rays and electrons with the interstellar medium.

$$\frac{\mathrm{d} N_{diffuse-galactic,\theta=0}(E_\gamma > E_{th})}{\mathrm{d}t\ \mathrm{d}A} = 5.1 \times 10^{-5} \left(\frac{\Delta\Omega}{1 sr}\right) \left(\frac{E_{th}}{1\ \mathrm{GeV}/c^2}\right)^{-1.7} \mathrm{cm}^{-2}\ \mathrm{s}^{-1} \quad (4.30)$$

Whereas this contribution can be neglected for ACT observations, it is the dominant background for space-borne experiments. This background however is not isotropic but concentrated around the galactic centre and in the galactic disc. We find that for both possible DM profiles for the main halo the signal-to-noise ratio is larger when instead of the galactic centre itself a ring OUTSIDE of the diffuse galactic component is observed. If we assume that the diffuse galactic background has dropped to the level of the extragalactic background outside of 30 degrees from the centre and 10 degrees of the galactic plane, the signal-to-noise ratio is expected to be about 12 times better than that for a 0.1 degree beam in the direction of the Galactic Centre (for an assumed NFW profile). This result is spectacular: the density profile of the DM halo in these regions is well established from the simulations and the prediction becomes *independent* of numerical uncertainties in the innermost structure of CDM haloes.

Finally, EGRET has also observed a diffuse extragalactic γ-ray background which is supposed to stem from unresolved sources (Fichtel et al. (1983); Ullio et al. (2002)).

$$\frac{\mathrm{d} N_{extra-galactic}(E_\gamma > E_{th})}{\mathrm{d}t\ \mathrm{d}A} = 1.2 \times 10^{-6} \left(\frac{\Delta\Omega}{1 sr}\right) \left(\frac{E_{th}}{1\ \mathrm{GeV}/c^2}\right)^{-1.1} \mathrm{cm}^{-2}\ \mathrm{s}^{-1} \quad (4.31)$$

The backgrounds taken into account for the different detectability estimates are summarised in Table 4.2.

4.8.3 Results

These results are shown in Figure 4.11. The solid curve gives our estimated lower limit on the cross-section for a 3-σ detection of the Galactic Centre in a 0.1 degree beam for a 250 hr observation with the planned ACT VERITAS. This particular calculation extrapolates to small radii using the NFW model of Figure 4.9 and the S/N is then maximised for the smallest resolved detection cell. The short-dashed curve is the corresponding limit for inward extrapolation using the SWTS model. In this case the S/N is maximised using a detection cell of radius 1.75 degrees, corresponding to the full field-of-view of VERITAS. The corresponding lower limit for detection of the brightest substructure in our six artificial skies, again for a 1.75 degree beam, is shown by the long dashed curve.

4.8 Detectability

| | ground-based | | space-born | |
	centre	subhalo	centre	subhalo
Hadronic	no	no	no	no
Electron-induced	yes	yes	no	no
Diffuse-galactic	yes	no	yes	no
Extra-galactic	yes	yes	yes	yes

Table 4.2: Summary of the different γ-ray backgrounds used for the detectability estimates.

We assume the inner density structure of this object to be correctly described by our SWTS model fit. A 250 hr observation may just be enough to detect the Galactic Centre, at least for a few of the plausible MSSM models. Detection of substructure appears out of reach unless our simulations have grossly underestimated the central concentration of subhaloes.

Figure 4.11 also shows cross-section limits for a 1 year exposure with the satellite telescope GLAST. The straight long-dashed line is the limit for detecting the brightest satellite, assuming this to be outside the region with strong diffuse Galactic emission and using a detection cell of radius 5 degrees corresponding to the the peak S/N-angle. The straight solid line gives the limit for detecting annihilation radiation in an annulus between 25 and 35 degrees from the Galactic Centre but excluding the region within 10 degrees of the Galactic Plane. We assume that the total diffuse Galactic emission in this region is zero. The results here are quite encouraging. The inner Galaxy should be detectable for most allowable MSSM parameters, while the brightest substructure is also detectable for many of them (for TO's implicitly adopted prior on the MSSM parameter space).

Whereas the field-of-view of GLAST covers almost a fifth of the full sky, the smaller field-of-view of VERITAS allows observation of only one object at a time. In addition, ACTs can only operate about 6 hours per night. For these reasons we consider exposure times of 1 year for GLAST and 250 hours for VERITAS to be large but feasible.

4. Dark Matter Annihilation

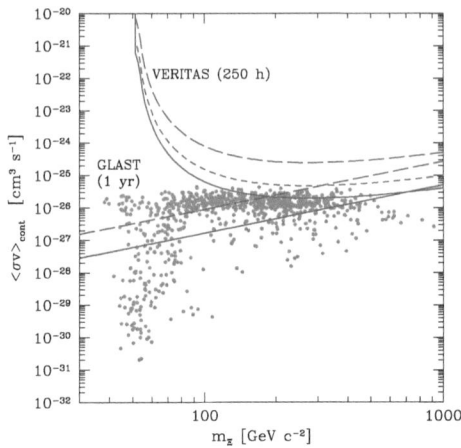

Figure 4.11: MSSM models of cosmological interest (dots) and 3-σ detection limits for VERITAS and GLAST. For VERITAS the limits are shown for a pointing at the centre of the Milky Way, assuming an NFW profile (solid) and an SWTS profile (short dashes). The shaded region gives estimated limits for GLAST for a larger area observation of the inner Galaxy which avoids regions of high contamination by diffuse Galactic emission. Limits for a pointing at the brightest high latitude subhalo are shown for both telescopes using long dashes. The brightest subhalo was chosen from the 6 artificial skies used in making Figure 4.9.

Table 4.3: Simplified telescope specifications

	A_{eff} [m^2]	E_{th} [GeV/c^2]	σ_θ [arcmin]
VERITAS	10^4	50	6
GLAST	0.8	0.02	6

4.9 Conclusions

We have directly estimated the γ-ray emissivity of the Milky Way's halo using high resolution simulations of its formation in a standard ΛCDM universe. A series of resimulations of the same DM halo at different mass resolution allows us to check explicitly for numerical convergence in our results. We find that the resolution limit of our largest simulation is almost an order of magnitude smaller than the half-light radius for the annihilation radiation, and that our estimates of the total flux are almost converged. We argue that the annihilation radiation from substructure within the Galactic halo is dominated by the most massive subhaloes, is concentrated in the outer halo, and is less in total than the radiation from the smooth inner halo. For the most massive subhaloes our convergence study indicates sufficient resolution in our best simulations to get robust estimates of their internal structure. An important result is that subhalo cores are less concentrated both than that of the main halo and than those of their progenitor haloes. This confirms earlier results by SWTS and Hayashi et al (2003) and apparently reflects the influence of tidal shocking on subhalo structure.

We find that 15% of the total flux in our highest resolution simulation is coming from gravitationally bound subhaloes and that no more than about 5% can be assigned to other small-scale density fluctuations. Some of our results do strongly rely on the density behaviour we infer from our simulations for the innermost regions both of the main halo and of the subhaloes. This subject is still controversial, although the most detailed convergence study to date agrees quite well with our results for the centre of the main halo (Power et al. 2003). As already noted our subhalo structure agrees well with that found by Hayashi et al (2003). The disagreement between our results and other theoretical work on annihilation luminosities (Calcaneo-Rodan & Moore 2000, Tasitsiomi & Olinto 2002, Taylor & Silk 2003) can be traced to the fact that the density profiles adopted in these papers are incompatible with those we measure in our simulations, particularly for subhaloes. It may be relevant that observational data on dwarf galaxies also speak in favour of dark matter density profiles with low concentrations or cores (de Blok et al. 2001b) although here again the situation is controversial.

To estimate the fluxes expected for deep integrations with upcoming experiments, it is necessary to extrapolate density profiles down to scales at least an order of magnitude below those where they can be reliably estimated in our simulations. This clearly introduces substantial uncertainty. Our results suggest that the central regions of the Galaxy will be intrinsically more luminous than the brightest substructure by a factor of at least ten, and apparently more luminous by a factor approaching one thousand. The angular scale associated with the central emission will be several tens of degrees while that associated with the substructure will be a few degrees. This suggests that it

4. Dark Matter Annihilation

may be worthwhile to investigate detection strategies which are sensitive to large-scale diffuse emission. Notice that since our results imply that the most apparently luminous subhaloes will also be among the most massive, it is likely that the brightest substructure source will be identified with one of the known satellites of the Milky Way. The closest of these is the Sagittarius dwarf at a distance of 24 kpc, but it may well be outshone by the LMC at a distance of 45 kpc. Both are sufficiently far that they will be much fainter (and smaller in angular size) than the main halo source which is centred only 8 kpc away.

Following TO and using DARKSUSY we checked for detectability of the inner Milky Way with VERITAS and GLAST, examples of ground- and space-based next-generation telescopes, respectively. If we extrapolate our simulated density profiles inwards using an NFW fit, VERITAS can probe into the parameter ranges in which a minimal supersymmetric extension of the Standard Model could provide a Dark Matter candidate with the observed cosmic density. Unfortunately, extrapolating inwards using our SWTS fit, which appears to provide a better description of our simulations, results in lower predicted fluxes, undetectable for VERITAS.

By searching for extended emission outside the central region where diffuse Galactic γ-ray emission is dominant GLAST can probe a large region of possible MSSM models. This result is based on the DM distribution in regions where the simulations have reliably converged, and so should be robust. It is *independent* of the exact structure of the DM in the innermost regions.

Our simulations suggest that the flux from the inner Galaxy will outshine the brightest substructure by a large factor. Nevertheless, for certain MSSM models some of the most massive substructure haloes might be detectable with GLAST. Clearly the most massive and nearest *known* satellite galaxies are the primary targets for observation.

Conclusions

In this work we have extensively used the resimulation technique to address aspects of structure formation.

In the first chapter we studied the influence of the local environment density of DM haloes on their properties. We ran a series of high-resolution simulations for a carefully selected representative region of a ΛCDM universe. The mass resolution reached in our largest simulation is the highest resolution obtained in a comparable volume so far. The simulation series has already been widely used by different groups to study the propagation of reionisation fronts, metal enrichment in the intergalactic medium through supernova driven winds, substructure in DM haloes, a new tessellation method, the distribution of satellites around large galaxies in the 2dF Survey, mass accretion histories of haloes and their concentration parameters, spin-spin correlation of DM haloes and predictions for observations of the Lyman-α forest.

We analysed the mass function of haloes in our simulations and compared it to the analytical prediction of Sheth et al. (2001b), extending the range over which this prediction is tested by two orders of magnitude. We find that their formula describes very well our results at $z = 0$. At higher redshifts, however, we measure mass functions that have significantly steeper slopes.

We proposed a new method of estimating the density around dark matter haloes that only takes into account those parts of the density field that do not belong to collapsed regions. We used this method and the one proposed by Lemson & Kauffmann (1999) and find that the concentration parameter of the halo density profiles as well as the spin parameter positively correlate with the local density. Haloes in denser environments are on average more concentrated and rotate faster than their low-density environment counterparts. We confirm the results of Lemson & Kauffmann (1999) finding no correlation of halo formation time, halo shape or the last major merging event with the environment.

In a second step we applied a semi-analytic model of galaxy formation to the results of our dark matter simulation series. We find the luminosity function of the model galaxies and the global star formation history to provide a good match of observations. We confirm the finding of other authors that the faint-end slope of the simulated luminosity function is steeper than those of the Sloan Digital Sky Survey and the 2dF Survey. Additional physics (i.e. photoheating) has to be included in the models.

Our galaxy sample nicely reproduces the slope of the Tully-Fisher relation. The normalisation can be brought into agreement with observations by assuming that the stellar component has a characteristic velocity of 85% of the virial velocity. Our simulations of chapter 2 suggest that this assumption is plausible.

We find a strong luminosity ratio and B-V dependence on environment, consistent with observations. The model also correctly predicts the trend of the colour-density relation found in observations and we obtain a bimodal distribution of galaxy colours.

In the second chapter we addressed the so-called "substructure problem" which is one of the two major challenges of the ΛCDM model of cosmology. We performed ultra-high resolution simulations of the assembly of a Milky Way type dark matter halo within its full cosmological context. We ran a series of four resimulations with increasing mass resolution, reaching in our largest simulation the highest resolution achieved so far.

We identified subhaloes within the main galaxy halo and proposed a new analytical fitting formula (SWTS) for their circular velocity curves. This formula provides a very good fit for the profiles of all of our subhaloes. We find that the corresponding density profiles are shallower in their very inner part than the universal profile found for haloes (Navarro, Frenk & White 1997). We verified the convergence of our results with object-by-object comparison between simulations of different mass resolution.

Our new profile also provides a better fit to the simulated Milky Way halo than the NFW profile does. If future simulations confirm this result, it might bring simulations in closer agreement with the observed density cores in galaxies and could thus solve the second major problem of the ΛCDM model.

Using observed stellar distributions of the Milky Way dwarf spheroidal galaxies, we predict their central velocity dispersions by placing them in our simulated potential wells. The observed kinematics exactly match those predicted by our simulations. In addition, we show that our simulations not only predict the central dispersion value but also the correct shape of the full velocity dispersion profile for Fornax and Draco, where such profiles have been observed.

We find that it is more appropriate to consider the observed kinematics of the Milky Way's satellites a "triumph" for the ΛCDM model than a problem.

If our findings are correct then the measured "tidal-radii" of dwarf spheroidals only mark the edge of the stellar distribution within a much larger dark matter halo. They have nothing to do with tidal effects. A second consequence is that the dark matter mass of dwarf galaxies is much larger than previously assumed. If this is true, star formation must be strongly suppressed in these systems. Furthermore, our results predict that many small substructure haloes devoid of stars are present in the Milky Way halo.

The fact that the structure of the subhaloes agrees with the observed satellites is evidence against dark matter with modified properties like warm dark matter or self-interacting dark matter.

In the third chapter we used our ultra-high resolution simulations to study the possible γ-ray signal from dark matter annihilation. If such a signal was detected, the nature of the dark matter, one of the most important questions of modern cosmology, would be known.

We used a Voronoi-tessellation density estimate and computed the annihilation signal from the centre of the Milky Way as well as from the subhaloes. We find that the γ-ray emission is rather extended. Half of the light comes from a smooth dark matter distribution emitted from a region of about 7 kpc around the halo centre. Subhaloes contribute with about 15% to the total flux and with about 5% to the total mass. We find that this annihilation signal from subhaloes is dominated by the largest objects. We studied the distribution of the subhaloes within the main halo and find that they are concentrated in the outskirts: Only one percent of the substructure haloes is closer than 80 kpc to the galactic centre.

One of the most probable candidates for the dark matter particle is the neutralino arising in minimally supersymmetric extensions of the standard model of particle physics (MSSM). Following Tasitsiomi & Olinto (2002) and using the package DARKSUSY, we estimated the detectability of the corresponding annihilation signal. If we extrapolate our simulated density profiles inwards using an NFW fit, VERITAS can probe into the parameter ranges where the MSSM model could provide a dark matter candidate with the observed cosmic relic density, Ω_{DM}. An inward extrapolation with the our proposed SWTS profile, which appears to provide a better description of the results of our simulations, predicts lower fluxes, undetectable for VERITAS.

For an observation outside the region around the galactic centre where the diffuse galactic γ-ray background is dominant GLAST can probe a large range of possible MSSM models. This result is based only on the well established structure of the DM distribution at relatively large distances from the galactic centre. It is independent of the exact density distribution in the innermost regions.

Our analysis shows that the flux from the galactic centre is likely to outshine the contribution from the brightest substructure by a large factor. Nevertheless, for certain MSSM models subhaloes might be detectable with GLAST, although detailed modelling of the best candidates, the closest of the known Milky Way dwarfs, is necessary.

Conclusions

Recent years have witnessed dramatic progress in cosmology. The theory of structure formation seems to be mature and the cosmological parameters are measured with unprecedented accuracy. We now know that we live in an accelerating Universe where about nine tenth of matter is dark. Upcoming or already operating experiments (SDSS, Glast, NGST, Planck, ...) will provide us with fascinating new data at a fast pace.

On the other hand, the power of supercomputers seems to increase without rest. When this PhD thesis was started, the supercomputer of the Max-Planck-Society was the 24th fastest computer in the world. The new supercomputer of the Max-Planck-Society is the 25th fastest computer in the world today. However, this new computer is more than 6 times more powerful than its predecessor. We live in exciting times.

List of Publications

Articles

Jeremy Blaizot, Bruno Guiderdoni, Julien E. G. Devriendt, François Bouchet, Steve Hatton and Felix Stoehr
GALICS III: Predicted properties for Lyman Break Galaxies at redshift 3,
MNRAS submitted, astro-ph/0310071, 2003

Felix Stoehr, Simon D. M. White, Giuseppe Tormen, Volker Springel and Naoki Yoshida
Dark matter annihilation in the Milky Way's halo,
MNRAS, volume 345, pages 1313-1322, 2003

Benedetta Ciardi, Felix Stoehr and Simon D. M. White,
Simulating inter galactic medium Reionization,
MNRAS, volume 343, pages 1101-1109, 2003

Felix Stoehr, Simon D. M. White, Giuseppe Tormen and Volker Springel,
The satellite population of the Milky Way in a ΛCDM universe,
MNRAS, volume 335, pages L84-L88, 2002

Naoki Yoshida, Felix Stoehr, Volker Springel and Simon D. M White,
Gas cooling in simulations of the formation of the galaxy population,
MNRAS, volume 335, pages 762-772, 2002

Proceedings

Simon D. M. White, Benedetta Ciardi, Piero Madau and Felix Stoehr
Mapping the reionization era through the 21CM line emission
Maps of the Cosmos, IAU. Symposium no. 216, 2003

List of Publications

Jeremy Blaizot, Bruno Guiderdoni, Julien E. G. Devriendt, François R. Bouchet, Steve Hatton and Felix Stoehr,
Optical/IR Properties of High-z Galaxies in the GalICS Model of Hierarchical Galaxy Formation,
SF2A-2002: Semaine de l'Astrophysique Française, 2002

Julien E. G. Devriendt, Stéphane Ninin, Bruno Guiderdoni, Stéphane Colombi, Steve Hatton, Felix Stoehr and Didier Vibert,
Galaxies Cooked the N-Body Plus S.A.M. Way,
XIXth Moriond Astrophysics Meeting,1999

Popular articles

Felix Stoehr and Simon D. M. White,
Ballet de galaxies,
Pour La Science, Dossier N° 38: La Gravitation, N° SAP 077638, 2003

Felix Stoehr and Simon D. M. White,
Ballett der Galaxien,
Sterne und Weltraum, Special 6: Gravitation, ISSN 1434-2057, 2001

Diploma Thesis

Felix Stoehr,
High Resolution Simulations of Structure Formation in Underdense Regions,
Technische Universität München, 1999

List of Figures

1.1	Contributions of the different forms of matter to the total matter density	20
1.2	Cooling functions	31
1.3	Simulation of reionisation using the M-series	34
2.1	Lagrangian region of M3	43
2.2	Eulerian region of M3	44
2.3	Powerspectrum	47
2.4	Mass in shells	51
2.5	FOF subtraction	52
2.6	Estimator comparison	53
2.7	Cumulative mass functions	55
2.8	Massfunctions	56
2.9	Profiles of most massive M3 halo	58
2.10	Concentration parameter	59
2.11	Averaged profiles for M3 haloes	60
2.12	Short-to-long axis ratio	62
2.13	Intermediate-to-long axis ratio	63
2.14	Short-to-long axis ratio in mass cuts	64
2.15	Spin parameter	65
2.16	Formation time	67
2.17	Formation time in mass slices	68
2.18	Redshift of the last major Merger	69
2.19	Luminosity Function	76
2.20	Tully-Fisher relation	77
2.21	Scaled Tully-Fisher relation	78
2.22	Madau-plot	79
2.23	Galaxy distribution in S3 and M3 at $z=5$	81
2.24	Galaxy distribution in S3 and M3 at $z=2$	82
2.25	Galaxy distribution in S3 and M3 at $z=0$	83

List of Figures

2.26 Galaxy distribution above DM halo masses of 10^9 $M_\odot/h$ in S3 and M3 at $z = 0$ 84
2.27 Bulge-to-total 86
2.28 B-V colour index 87

3.1 G-2 profiles 97
3.2 Radial velocity and mass growth 98
3.3 G-2 merginghistory 99
3.4 Startingredshift 103
3.5 GA3n profiles 108
3.6 GA3n mass growth 109
3.7 GA3n merginghistory 110
3.8 GA3n, GA2n, GA1n and GA0n at $z = 0$ 111
3.9 Assembly of GA3n 112
3.10 Bound and unbound halos: (r_{max}, v_{max}) 117
3.11 Accretion of the subhaloes 118
3.12 Change of r_{max} and v_{max} values of infalling subhaloes 120
3.13 Relative change of v_{max} and r_{max} of infalling subhaloes 121
3.14 Mass-loss fraction of infalling subhaloes 122
3.15 Circular velocity curves for GA3n subhaloes 124
3.16 Hayashi et al. (2003) circular velocity curves 125
3.17 Bound vs. unbound subhalo profiles 126
3.18 a-values for the simulations GA1n, GA2n and GA3n 127
3.19 (r_{max}, v_{max}) values for the simulations GA1n, GA2n and GA3n 128
3.20 r_{max}, v_{max} and a 130
3.21 Velocity dispersion profiles for Fornax and Draco 132
3.22 Line of sight projection 135
3.23 Predicted velocity dispersion profiles 139
3.24 Predicted velocity dispersion profiles using Hayashi et al. (2003) 140

4.1 GA3n annihilation luminosity 150
4.2 Circular velocity curves for the simulations GA0n, GA1n, GA2n and GA3n 151
4.3 Subhalo mass function for GA3n and S4 154
4.4 Cumulative number of subhaloes in the four simulations GA0n, GA1n, GA2n and GA3n 155
4.5 Voronoi tessellation 156
4.6 Annihilation luminosity 157
4.7 Luminosity contributions 159

4.8	Cumulative luminosity	162
4.9	Surface brightness and signal-to-noise	164
4.10	Relic densities and line of sight flux	168
4.11	Detectability	172

List of Figures

List of Tables

1.1	Cosmological parameters	23
2.1	Simulation parameters	46
2.2	Parameters of the M-series simulations	48
2.3	ZIC parameters	49
2.4	Free parameters of the semi-analytic model	74
3.1	Galaxy candidates	96
3.2	Coordinates of the halo centres	100
3.3	Local group project	102
3.4	Parameters of the GA-series	105
3.5	Resimulated haloes	106
3.6	Dwarf galaxy data	131
3.7	Number of simulated subhalo profiles that predict a *larger* central velocity dispersion than that observed	137
4.1	MSSM parameters	167
4.2	γ-ray backgrounds	171
4.3	Simplified telescope specifications	172

List of Tables

Units and Constants

speed of light	c =	299 792 458 m s^{-1}
year (siderial)	yr =	31 558 149.8 s $\approx$
	$\approx$	$\pi 10^7 s$
mean day (siderial)	d =	23^h 56^m $04.^s 09053$
parsec	pc =	3.085 677 580 7(4) 10^{16} m =
	=	3.262 ly
astronomical unit	au =	149 597 870 660 (20) m
solar distance from the galactic centre	$R_\circ$ =	8.0(5) kpc
solar velocity around the galactic centre	$\theta_\circ$ =	220(20) km s^{-1}
sun mass	$M_\odot$ =	1.988 92(25) 10^{30} kg
earth mass	$M_\otimes$ =	5.974(9) $\times 10^{24}$ kg
gravitational constant	G_N =	6.672 59(85) 10^{-11} m^3 kg^{-1} s^{-2} =
	=	4.30091 10^{-9} Mpc km^2 s^{-2} $M_\odot^{-1}$
Hubble constant	H_0 =	100 h_0 km s^{-1} Mpc^{-1} =
	=	h_0 (9.778 13 Gyr)$^{-1}$
Hubble time	H_0^{-1} =	9.776 10^9 h_0^{-1} yr
Hubble radius	c/H_0 =	2997.9 h_0^{-1} Mpc
critical density	ρ_c =	$3H_0^2/8\pi G_N$ =
	=	2.775 366 27 10^{11} h_0^2 $M_\odot$ / Mpc^3 =
	=	1.0539(16) $\times 10^{-5}$ h^2 GeV/c^2 cm^{-3}
local halo density	ρ_{halo} =	2-13 $\times 10^{28}$ kg/m^3 =
	=	0.1-0.7 GeV/c^2 cm^{-3}
CMB temperature	T_0 =	2.725(1) K

Units and Constants

Bibliography

Alpher, R. A., Bethe, H., Gamov, G. (1948), ., Phys. Rev. Letters., 70, 73(803)   19

Baltz, E. A., Briot, C., Salati, C., Taillet, R., Silk, J. (2000), *Detection of neutralino annihilation photons from external galaxies*, Physical Review, 61   144, 169

Barnes, J., Hut, P., Goodman, J. (1986), *Dynamical instabilities in spherical stellar systems*, ApJ , 300, 112   36

Baumann, D., Cooray, A., Kamionkowski, M. (2003), *Small-scale cosmic microwave background polarization from reionization.*, New Astronomy, 8, 565   92

Bell, E. F., Wolf, C., Meisenheimer, K., Rix, H., Borch, A., Dye, S., Kleinheinrich, M., McIntosh, D. H. (2003), *Over 5000 Distant Early-Type Galaxies in COMBO-17: a Red Sequence and its Evolution since $z\~1$*, ArXiv Astrophysics e-prints   85

Benson, A. J., Bower, R. G., Frenk, C. S., Lacey, C. G., Baugh, C. M., Cole, S. (2003), *What Shapes the Luminosity Function of Galaxies?*, ArXiv Astrophysics e-prints   73, 75

Benson, A. J., Frenk, C. S., Lacey, C. G., Baugh, C. M., Cole, S. (2002), *The effects of photoionization on galaxy formation - II. Satellite galaxies in the Local Group*, MNRAS , 333, 177   92

Bergström, L., Edsjö, J., Gondolo, P., Ullio, P. (1999), *Clumpy Neutralino Dark Matter*, Physical Review, 59   152, 153, 161

Bergström, L., Ullio, P., Buckley, J. H. (1998), *Observability of gamma rays from dark matter neutralino annihilations in the Milky Way halo*, Astroparticle Physics, 9, 137   144, 152, 167, 169

Bertschinger, E. (1995), *COSMICS: Cosmological Initial Conditions and Microwave Anisotropy Codes*, astro-ph/9506070   101

Bibliography

Bertschinger, E. (2001), *Multiscale Gaussian Random Fields and Their Application to Cosmological Simulations*, ApJS , 137, 1   42

Binney, J., Tremaine, S. (1987), *Galactic dynamics*, Princeton, NJ, Princeton University Press, 1987, 747 p.   133, 134

Blanton, M. R., Dalcanton, J., Eisenstein, D., Loveday, J., Strauss, M. A., SubbaRao, M., Weinberg, D. H., Anderson, J. E., Annis, J., Bahcall, N. A., Bernardi, M., Brinkmann, J., Brunner, R. J., Burles, S., Carey, L., Castander, F. J., Connolly, A. J., Csabai, I., Doi, M., Finkbeiner, D., Friedman, S., Frieman, J. A., Fukugita, M., Gunn, J. E., Hennessy, G. S., Hindsley, R. B., Hogg, D. W., Ichikawa, T., Ivezić, Ž., Kent, S., Knapp, G. R., Lamb, D. Q., Leger, R. F., Long, D. C., Lupton, R. H., McKay, T. A., Meiksin, A., Merelli, A., Munn, J. A., Narayanan, V., Newcomb, M., Nichol, R. C., Okamura, S., Owen, R., Pier, J. R., Pope, A., Postman, M., Quinn, T., Rockosi, C. M., Schlegel, D. J., Schneider, D. P., Shimasaku, K., Siegmund, W. A., Smee, S., Snir, Y., Stoughton, C., Stubbs, C., Szalay, A. S., Szokoly, G. P., Thakar, A. R., Tremonti, C., Tucker, D. L., Uomoto, A., Vanden Berk, D., Vogeley, M. S., Waddell, P., Yanny, B., Yasuda, N., York, D. G. (2001), *The Luminosity Function of Galaxies in SDSS Commissioning Data*, AJ , 121, 2358   75

Blumenthal, G. R., Faber, S. M., Primack, J. R., Rees, M. J. (1984), *Formation of galaxies and large-scale structure with cold dark matter*, Nature , 311, 517   92

Bode, P., Ostriker, J. P., Turok, N. (2001a), *Halo Formation in Warm Dark Matter Models*, ApJ , 556, 93   94

Bode, P., Ostriker, J. P., Turok, N. (2001b), *Halo Formation in Warm Dark Matter Models*, ApJ , 556, 93   142

Bond, J. R., Cole, S., Efstathiou, G., Kaiser, N. (1991), *Excursion set mass functions for hierarchical Gaussian fluctuations*, ApJ , 379, 440   28, 49, 88

Bond, J. R., Efstathiou, G. (1987), *The statistics of cosmic background radiation fluctuations*, MNRAS , 226, 655   24, 46, 47, 100

Bruzual A., G., Charlot, S. (1993), *Spectral evolution of stellar populations using isochrone synthesis*, ApJ , 405, 538   32, 74

Bullock, J. S., Kolatt, T. S., Sigad, Y., Somerville, R. S., Kravtsov, A. V., Klypin, A. A., Primack, J. R., Dekel, A. (2001), *Profiles of dark haloes: evolution, scatter and environment*, MNRAS , 321, 559   40, 160

Bullock, J. S., Kravtsov, A. V., Weinberg, D. H. (2000), *Reionization and the Abundance of Galactic Satellites*, ApJ , 539, 517   92

Burkert, A. (1995), *The Structure of Dark Matter Halos in Dwarf Galaxies*, ApJ , 447, L25+   93

Calcanéo-Roldán, C., Moore, B. (2000), *The surface brightness of dark matter: unique signatures of neutralino annihilation in the Galactic halo*, Physical Review, 62   148, 152, 153

Chiba, M. (2002), *Probing Dark Matter Substructure in Lens Galaxies*, ApJ , 565, 17   116, 142

Ciardi, B., Ferrara, A., White, S. D. M. (2003a), *Early Reionization by the First Galaxies*, ArXiv Astrophysics e-prints, 2451–+   33, 66

Ciardi, B., Stoehr, F., White, S. D. M. (2003b), *Simulating IGM Reionization*, ArXiv Astrophysics e-prints, 1293–+   33, 34, 66, 79

Cole, S., Lacey, C. (1996), *The structure of dark matter haloes in hierarchical clustering models*, MNRAS , 281, 716+   27, 54

Cole, S., Lacey, C. G., Baugh, C. M., Frenk, C. S. (2000), *Hierarchical galaxy formation*, MNRAS , 319, 168   37, 49, 74, 123

Couchman, H. M. P., Rees, M. J. (1986), *Pregalactic evolution in cosmologies with cold dark matter*, MNRAS , 221, 53   92

Cowie, L. L., Songaila, A., Barger, A. J. (1999), *Evidence for a Gradual Decline in the Universal Rest-Frame Ultraviolet Luminosity Density for Z ¡ 1*, AJ , 118, 603   79

Dalal, N., Kochanek, C. S. (2002), *Direct Detection of Cold Dark Matter Substructure*, ApJ , 572, 25   116, 142

Davis, M., Efstathiou, G., Frenk, C. S., White, S. D. M. (1985), *The evolution of large-scale structure in a universe dominated by cold dark atter*, ApJ , 292, 371   50, 92

de Blok, W. J. G., McGaugh, S. S., Rubin, V. C. (2001a), *High-Resolution Rotation Curves of Low Surface Brightness Galaxies. II. Mass Models*, AJ , 122, 2396   94

de Blok, W. J. G., McGaugh, S. S., Rubin, V. C. (2001b), *High-Resolution Rotation Curves of Low Surface Brightness Galaxies. II. Mass Models*, AJ , 122, 2396   115, 173

Bibliography

Dekel, A., Arad, I., Devor, J., Birnboim, Y. (2002), *Dark-Halo Cusp: Asymptotic Convergence*, in *astro-ph/0205448*   107, 152

Devriendt, J., Ninin, S., Guiderdoni, B., Colombi, S., Hatton, S., Stöhr, F., Vibert, D. (1999), *Galaxies Cooked the N-Body Plus S.A.M. Way*, in to appear in the proceedings of the XIXth Moriond Astrophysics Meeting   37

Doroshkevich, A. G., Naselsky, I. P., Naselsky, P. D., Novikov, I. D. (2003), *Ionization History of the Cosmic Plasma in the Light of the Recent Cosmic Background Imager and Future Planck Data*, ApJ , 586, 709   92

Doroshkevich, A. G., Zel'dovich, Y. B., Novikov, I. D. (1967), *The Origin of Galaxies in an Expanding Universe.*, Soviet Astronomy, 11, 233   92

Dressler, A. (1980), *A catalog of morphological types in 55 rich clusters of galaxies*, ApJS , 42, 565   17, 40, 85

Dubinski, J., Carlberg, R. G. (1991), *The structure of cold dark matter halos*, ApJ , 378, 496   57

Efstathiou, G. (1992), *Suppressing the formation of dwarf galaxies via photoionization*, MNRAS , 256, 43P   92

Efstathiou, G., Davis, M., White, S. D. M., Frenk, C. S. (1985), *Numerical techniques for large cosmological N-body simulations*, ApJS , 57, 241   33, 36

Efstathiou, G., Eastwood, J. W. (1981), *On the clustering of particles in an expanding universe*, MNRAS , 194, 503   36

Eke, V. R., Cole, S., Frenk, C. S. (1996), *Cluster evolution as a diagnostic for Omega*, MNRAS , 282, 263   24, 28

Fichtel, C. E., Bertsch, D. L., Hartman, R. C., Kniffen, D. A., Thompson, D. J., Hofstadter, R., Hughes, E. B., Campbell-Finman, L. E., Pinkau, K., Mayer-Hasselwander, H. (1983), *EGRET - The high energy gamma ray telescope for NASA's Gamma Ray Observatory*, International Cosmic Ray Conference, 18th, Bangalore, India, August 22-September 3, 1983, Conference Papers. Volume 8 (A85-22801 09-93). Bombay, Tata Institute of Fundamental Research, 1983, p. 19-22., 8, 19   170

Firmani, C., D'Onghia, E., Chincarini, G., Hernández, X., Avila-Reese, V. (2001), *Constraints on dark matter physics from dwarf galaxies through galaxy cluster haloes*, MNRAS , 321, 713   93

Flores, R. A., Primack, J. R. (1994), *Observational and theoretical constraints on singular dark matter halos*, ApJ , 427, L1   93

Folkes, S., Ronen, S., Price, I., Lahav, O., Colless, M., Maddox, S., Deeley, K., Glazebrook, K., Bland-Hawthorn, J., Cannon, R., Cole, S., Collins, C., Couch, W., Driver, S. P., Dalton, G., Efstathiou, G., Ellis, R. S., Frenk, C. S., Kaiser, N., Lewis, I., Lumsden, S., Peacock, J., Peterson, B. A., Sutherland, W., Taylor, K. (1999), *The 2dF Galaxy Redshift Survey: spectral types and luminosity functions*, MNRAS , 308, 459   75, 76

Font, A. S., Navarro, J. F., Stadel, J., Quinn, T. (2001), *Halo Substructure and Disk Heating in a Λ Cold Dark Matter Universe*, ApJ , 563, L1   142, 153

Freedman, W. L., Madore, B. F., Gibson, B. K., Ferrarese, L., Kelson, D. D., Sakai, S., Mould, J. R., Kennicutt, R. C., Ford, H. C., Graham, J. A., Huchra, J. P., Hughes, S. M. G., Illingworth, G. D., Macri, L. M., Stetson, P. B. (2001), *Final Results from the Hubble Space Telescope Key Project to Measure the Hubble Constant*, ApJ , 553, 47   22, 23

Frenk, C. S., White, S. D. M., Davis, M., Efstathiou, G. (1988), *The formation of dark halos in a universe dominated by cold dark matter*, ApJ , 327, 507   57

Gamov, G. (1946), ., Phys. Rev., 70, 572   19

G.C. McVittie, N. Y. M., editor (1962), *In: Problems of Extragalactic Research*   17

Ghigna, S., Moore, B., Governato, F., Lake, G., Quinn, T., Stadel, J. (1998), *Dark matter haloes within clusters*, MNRAS , 300, 146   93

Ghigna, S., Moore, B., Governato, F., Lake, G., Quinn, T., Stadel, J. (2000), *Density Profiles and Substructure of Dark Matter Halos: Converging Results at Ultra-High Numerical Resolution*, ApJ , 544, 616   28, 40, 93, 114, 123

Gilmore, G., Wyse, R. F. G., Norris, J. E. (2002), *Deciphering the Last Major Invasion of the Milky Way*, ApJ , 574, L39   95

Giovanelli, R., Haynes, M. P., Herter, T., Vogt, N. P., da Costa, L. N., Freudling, W., Salzer, J. J., Wegner, G. (1997), *The I Band Tully-Fisher Relation for Cluster Galaxies: a Template Relation, its Scatter and Bias Corrections.*, AJ , 113, 53   75, 77

Goto, T., Okamura, S., McKay, T. A., Bahcall, N. A., Annis, J., Bernard, M., Brinkmann, J., Gómez, P. L., Hansen, S., Kim, R. S. J., Sekiguchi, M., Sheth, R. K.

Bibliography

(2002), *Composite Luminosity Functions Based on the Sloan Digital Sky Survey "Cut and Enhance" Galaxy Cluster Catalog*, PASJ , 54, 515   76

Gronwall, C. (1998), *The Evolution of Faint Field Galaxies: Implications from the Hubble Deep Field*, in *ASP Conf. Ser. 146: The Young Universe: Galaxy Formation and Evolution at Intermediate and High Redshift*, 96–+   79

Gunn, J. E., Peterson, B. A. (1965), *On the Density of Neutral Hydrogen in Intergalactic Space.*, ApJ , 142, 1633   33

Guth, A. H. (1981), ., Phys. Rev. D23, 347, 17   24

Hayashi, E., Navarro, J. F., Taylor, J. E., Stadel, J., Quinn, T. (2003), *The Structural Evolution of Substructure*, ApJ , 584, 541   123, 125, 126, 136, 138, 140, 141, 153, 154, 158, 182

Heavens, A., Peacock, J. (1988), *Tidal torques and local density maxima*, MNRAS , 232, 339   88

Helmi, A., White, S. (2000), *Halo Debris Streams as Relicts from the Formation of the Milky Way*, in *Astronomische Gesellschaft Meeting Abstracts*, 20–24   141

Helmi, A., White, S. D., Springel, V. (2002), *The phase-space structure of a dark-matter halo: Implications for dark-matter direct detection experiments*, Phys. Rev. D , 66, 63502   142, 153

Helmi, A., White, S. D. M. (1999), *Building up the stellar halo of the Galaxy*, MNRAS , 307, 495   95

Helmi, A., White, S. D. M. (2001), *Simple dynamical models of the Sagittarius dwarf galaxy*, MNRAS , 323, 529   131

Hernquist, L. (1990), *An analytical model for spherical galaxies and bulges*, ApJ , 356, 359   107, 152

Hoyle, F. (1949), *Problems of Cosmical Aerodynamics*, Dayton, Ohio   61

Hu, E. M., Cowie, L. L., McMahon, R. G., Capak, P., Iwamuro, F., Kneib, J.-P., Maihara, T., Motohara, K. (2002), *A Redshift z=6.56 Galaxy behind the Cluster Abell 370*, ApJ , 568, L75   17

Hubble, E. (1929), ., Proc. N.A.S., 15, 168   17

Hubble, E., Humason, M. . A. J. (1931), ., ApJ   17

Hut, P., White, S. D. M. (1984), *Can a neutrino-dominated Universe be rejected?*, Nature , 310, 637   21, 145

Ibata, R. A., Gilmore, G., Irwin, M. J. (1994), *A Dwarf Satellite Galaxy in Sagittarius*, Nature , 370, 194   129

Irwin, M. J., Bunclark, P. S., Bridgeland, M. T., McMahon, R. G. (1990), *A new satellite galaxy of the Milky Way in the constellation of Sextans*, MNRAS , 244, 16P   129

Jing, Y., Suto, Y. (1999), *Density Profiles of Dark Matter Halos are not Universal*, astro-ph/9909478   40

Jing, Y. P. (1999), *The density profile of equilibrium and non-equilibrium dark matter halos*, astro-ph/9901340   40

Jing, Y. P., Suto, Y. (2002), *Triaxial Modeling of Halo Density Profiles with High-Resolution N-Body Simulations*, ApJ , 574, 538   57

Johnston, K. V., Spergel, D. N., Haydn, C. (2002), *How Lumpy Is the Milky Way's Dark Matter Halo?*, ApJ , 570, 656   142

Kaplinghat, M., Chu, M., Haiman, Z., Holder, G. P., Knox, L., Skordis, C. (2003), *Probing the Reionization History of the Universe using the Cosmic Microwave Background Polarization*, ApJ , 583, 24   92

Katz, N., Weinberg, D. H., Hernquist, L. (1996), *Cosmological Simulations with TreeSPH*, ApJS , 105, 19   31

Kauffmann, G., Charlot, S. (1998), *Chemical enrichment and the origin of the colour-magnitude relation of elliptical galaxies in a hierarchical merger model*, MNRAS , 294, 705   85

Kauffmann, G., Colberg, J. M., Diaferio, A., White, S. D. M. (1999), *Clustering of galaxies in a hierarchical universe - I. Methods and results at z=0*, MNRAS , 303, 188   37, 50, 70, 72, 73, 74, 75

Kauffmann, G., Nusser, A., Steinmetz, M. (1997), *Galaxy formation and large-scale bias*, MNRAS , 286, 795   37

Kauffmann, G., White, S., Guiderdoni, B. (1993), *The formation and evolution of galaxies within merging dark matter haloes*, MNRAS , 264, 201+   37, 74, 92, 123

Kauffmann, G., White, S. D. M. (1992), *The observational properties of an Omega = 0.2 cold dark matter universe*, MNRAS , 258, 511   92

Bibliography

King, I. (1962), *The structure of star clusters. I. an empirical density law*, AJ , 67, 471
129

Klapdor-Kleingrothaus, H., Zuber, K. (1997), *Teilchenastrophysik*, Teubner 147

Klessen, R., Grebel, E., Harbeck, D. (2003), *Draco – A Failure of the Tidal Model*, in astro-ph/0302287 138

Kleyna, J. T., Wilkinson, M. I., Evans, N. W., Gilmore, G. (2001), *First Clear Signature of an Extended Dark Matter Halo in the Draco Dwarf Spheroidal*, ApJ , 563, L115
131, 132, 139

Klypin, A., Kravtsov, A. V., Valenzuela, O., Prada, F. (1999), *Where Are the Missing Galactic Satellites?*, ApJ , 52, 82 12, 13, 40, 93, 94, 138, 148, 153

Kochanek, C. S. (2003), *Gravitational Lens Time Delays in Cold Dark Matter*, ApJ , 583, 49 142

Kodama, T., Smail, I., Nakata, F., Okamura, S., Bower, R. G. (2001), *The Transformation of Galaxies within the Large-Scale Structure around a $z=0.41$ Cluster*, ApJ , 562, L9 85

Lacey, C., Cole, S. (1993), *Merger rates in hierarchical models of galaxy formation*, MNRAS , 262, 627 52, 66

Lake, G. (1990), *Detectability of gamma-rays from clumps of dark matter*, Nature , 346, 39+ 161

Lamb, D. Q., Haiman, Z. (2003), *Gamma-Ray Bursts a Probe of the Epoch of Reionization*, AAS/High Energy Astrophysics Division, 35, 0 92

Lanzoni, B., Ciotti, L. (2003), *Projection effects on the FP thickness: a Monte-Carlo exploration*, ArXiv Astrophysics e-prints 40

Lemson, G., Kauffmann, G. (1999), *Environmental influences on dark matter haloes and consequences for the galaxies within them*, MNRAS , 302, 111 50, 61, 66, 88, 175

Madau, P., Ferguson, H. C., Dickinson, M. E., Giavalisco, M., Steidel, C. C., Fruchter, A. (1996), *High-redshift galaxies in the Hubble Deep Field: colour selection and star formation history to $z\sim4$*, MNRAS , 283, 1388 76

Madgwick, D. S., Hawkins, E., Lahav, O., Maddox, S., Norberg, P., Peacock, J., Baldry, I. K., Baugh, C. M., Bland-Hawthorn, J., Bridges, T., Cannon, R., Cole, S., Colless,

M., Collins, C., Couch, W., Dalton, G., De Propris, R., Driver, S. P., Efstathiou, G., Ellis, R. S., Frenk, C. S., Glazebrook, K., Jackson, C., Lewis, I., Lumsden, S., Peterson, B. A., Sutherland, W., Taylor, K. (2003), *The 2dF Galaxy Redshift Survey: galaxy clustering per spectral type*, ArXiv Astrophysics e-prints 85

Majewski, S. R., Ostheimer, J. C., Kunkel, W. E., Patterson, R. J. (2000), *Exploring Halo Substructure with Giant Stars. I. Survey Description and Calibration of the Photometric Search Technique*, AJ , 120, 2550 141

Mao, S., Schneider, P. (1998), *Evidence for substructure in lens galaxies?*, MNRAS , 295, 587 142

Mateo, M. (1994), *Dark Matter in Dwarf Spheroidal Galaxies: Observational Constraints*, in *Dwarf Galaxies*, 309–+ 129

Mateo, M. (1997), *The Kinematics of Dwarf Spheroidal Galaxies*, in *ASP Conf. Ser. 116: The Nature of Elliptical Galaxies; 2nd Stromlo Symposium*, 259–+ 131, 132, 139

Mateo, M. L. (1998), *Dwarf Galaxies of the Local Group*, ARA&A , 36, 435 129, 131, 139, 141

Mathis, H., Lemson, G., Springel, V., Kauffmann, G., White, S. D. M., Eldar, A., Dekel, A. (2002), *Simulating the formation of the local galaxy population*, MNRAS , 333, 739 40, 75

Mathis, H., White, S. D. M. (2002), *Voids in the simulated local Universe*, MNRAS , 337, 1193 80

McGaugh, S. S., de Blok, W. J. G. (1998), *Testing the Dark Matter Hypothesis with Low Surface Brightness Galaxies and Other Evidence*, ApJ , 499, 41 93, 115

Merritt, D., Milosavljević, M., Verde, L., Jimenez, R. (2002), *Dark Matter Spikes and Annihilation Radiation from the Galactic Center*, Physical Review Letters, 88, 191301 144

Metcalf, R. B., Madau, P. (2001), *Compound Gravitational Lensing as a Probe of Dark Matter Substructure within Galaxy Halos*, ApJ , 563, 9 153

Metcalfe, N., Fong, R., Shanks, T. (1995), *CCD galaxy photometry and the calibration of photographic surveys*, MNRAS , 274, 769 75

Bibliography

Milgrom, M. (1983), *A Modification of the Newtonian Dynamics - Implications for Galaxy Systems*, ApJ, 270, 384   19

Mo, H. J., White, S. D. M. (1996), *An analytic model for the spatial clustering of dark matter haloes*, MNRAS, 282, 347   37

Moore, B. (1994), *Evidence against Dissipationless Dark Matter from Observations of Galaxy Haloes*, Nature, 370, 629   93

Moore, B., Ghigna, S., Governato, F., Lake, G., Quinn, T., Stadel, J., Tozzi, P. (1999), *Dark Matter Substructure within Galactic Halos*, ApJ, 524, L19   12, 13, 40, 114, 123, 138, 148, 153

Moore, B., Governato, F., Quinn, T., Stadel, J., Lake, G. (1998), *Resolving the Structure of Cold Dark Matter Halos*, ApJ, 499, L5   28, 93, 94, 114

Moore, B., Katz, N., Lake, G. (1996), *On the Destruction and Overmerging of Dark Halos in Dissipationless N-Body Simulations*, ApJ, 457, 455   93

Navarro, J. F., Frenk, C. S., White, S. D. M. (1995a), *The assembly of galaxies in a hierarchically clustering universe*, MNRAS, 275, 56   107

Navarro, J. F., Frenk, C. S., White, S. D. M. (1995b), *Simulations of X-ray clusters*, MNRAS, 275, 720   40, 107

Navarro, J. F., Frenk, C. S., White, S. D. M. (1996), *The Structure of Cold Dark Matter Halos*, ApJ, 462, 563+   107

Navarro, J. F., Frenk, C. S., White, S. D. M. (1997), *A Universal Density Profile from Hierarchical Clustering*, ApJ, 490, 493+   27, 40, 54, 93, 160, 176

Nuss, E., Moultaka, G., Falvard, A., Giraud, E., Jacholkowska, A., Jedamzik, K., Lavalle, J., Piron, F., Sapinski, M., Salati, P., Taillet, R. (2002), *Detecting supersymmetric dark matter in M31 with CELESTE ?*, ArXiv Astrophysics e-prints   144

Nusser, A., Sheth, R. K. (1999), *Mass growth and density profiles of dark matter haloes in hierarchical clustering*, MNRAS, 303, 685   107, 152

Olszewski, E. W. (1998), *Internal Kinematics of Dwarf Spheroidal Galaxies*, in *ASP Conf. Ser. 136: Galactic Halos*, 70–+   129

Peacock, J. A. (1993), *Physical Cosmology*, Princeton University Press   18, 21, 25, 27, 28, 29, 61

Peacock, J. A., Dodds, S. J. (1996), *Non-linear evolution of cosmological power spectra*, MNRAS , 280, L19   27

Peccei, R. D., Quinn, H. R. (1977), *CP conservation in the presence of pseudoparticles*, Physical Review Letters, 38, 1440   145

Peebles, P. J. E. (1969), *Origin of the Angular Momentum of Galaxies*, ApJ , 155, 393   61

Peebles, P. J. E. (1980), *The Large-Scale Structure of the Universe*, Princeton   25, 27, 107, 152

Penzias, A. A., Wilson, R. W. (1965), *A Measurement of Excess Antenna Temperature at 4080 Mc/s.*, ApJ , 142, 419   19

Power, C., Navarro, J. F., Jenkins, A., Frenk, C. S., White, S. D. M., Springel, V., Stadel, J., Quinn, T. (2003), *The inner structure of ΛCDM haloes - I. A numerical convergence study*, MNRAS , 338, 14   28, 40, 101, 102, 114, 115, 148, 152, 173

Press, W. H., Schechter, P. (1974), *FORMATION OF GALAXIES AND CLUSTERS OF GALAXIES BY SELF-SIMILAR GRAVITATIONAL CONDENSATION*, ApJ , 187, 425   28, 49

Sanders, R. H., McGaugh, S. S. (2002), *Modified Newtonian Dynamics as an Alternative to Dark Matter*, ARA&A , 40, 263   19

Seljak, U., Zaldarriaga, M. (1996), *A Line-of-Sight Integration Approach to Cosmic Microwave Background Anisotropies*, ApJ , 469, 437   24, 46, 100

Sheth, R. K., Mo, H. J., Tormen, G. (2001a), *Ellipsoidal collapse and an improved model for the number and spatial distribution of dark matter haloes*, MNRAS , 323, 1   28

Sheth, R. K., Mo, H. J., Tormen, G. (2001b), *Ellipsoidal collapse and an improved model for the number and spatial distribution of dark matter haloes*, MNRAS , 323, 1   28, 49, 54, 56, 88, 92, 103, 175

Somerville, R. S., Primack, J. R., Faber, S. M. (2001), *The nature of high-redshift galaxies*, MNRAS , 320, 504   37, 75, 76, 79

Spergel, D. N., Steinhardt, P. J. (2000), *Observational Evidence for Self-Interacting Cold Dark Matter*, Physical Review Letters, Volume 84, Issue 17, April 24, 2000, pp.3760-3763, 84, 3760   94, 142

Spergel, D. N., Verde, L., Peiris, H. V., Komatsu, E., Nolta, M. R., Bennett, C. L., Halpern, M., Hinshaw, G., Jarosik, N., Kogut, A., Limon, M., Meyer, S. S., Page, L., Tucker, G. S., Weiland, J. L., Wollack, E., Wright, E. L. (2003), *First Year Wilkinson Microwave Anisotropy Probe (WMAP) Observations: Determination of Cosmological Parameters*, ArXiv Astrophysics e-prints, 2209–+ 19, 20, 21, 23, 24, 94, 167, 168

Springel, V., Hernquist, L. (2003), *Cosmological smoothed particle hydrodynamics simulations: a hybrid multiphase model for star formation*, MNRAS , 339, 289 30, 38

Springel, V., White, S. D. M. (1999), *Tidal tails in cold dark matter cosmologies*, MNRAS , 307, 162 33

Springel, V., White, S. D. M., Tormen, G., Kauffmann, G. (2001a), *Populating a cluster of galaxies - I. Results at z=0.*, MNRAS, 328, 726 37, 40, 46, 48, 70, 73, 74, 75, 79, 80, 115, 153

Springel, V., Yoshida, N., White, S. D. M. (2001b), *GADGET: A code for collisionless and gasdynamical cosmological simulations*, New Astronomy, 6, 79 36

Stanimirovic, S. (2000), *The complex nature of the ISM in the SMC: an HI and IR study*, Ph.D. thesis, Univ. Western Sydney Nepean, http://www.naic.edu/ sstanimi/thesis.html 129

Steidel, C. C., Adelberger, K. L., Giavalisco, M., Dickinson, M., Pettini, M. (1999), *Lyman-Break Galaxies at z ¿ 4 and the Evolution of the Ultraviolet Luminosity Density at High Redshift*, ApJ , 519, 1 79

Steinmetz, M. (1999), *Numerical Simulations of Galaxy Formation*, Ap&SS , 269, 513 40

Steinmetz, M., Bartelmann, M. (1995), *On the spin parameter of dark-matter haloes*, MNRAS , 272, 570 61, 88

Stoehr, F. (1999), *High Resolution Simulations of Structure Formation in Underdense Regions*, Master's thesis, Technische Universität München, http://www.iap.fr/users/stohr 40, 41, 52, 54, 95, 96, 100

Stoehr, F., White, S. D. M., Tormen, G., Springel, V. (2002), *The satellite population of the Milky Way in a ΛCDM universe*, MNRAS , 335, L84 105, 115, 126

Subramanian, K., Cen, R., Ostriker, J. P. (2000), *The Structure of Dark Matter Halos in Hierarchical Clustering Theories*, ApJ , 538, 528 107, 152

Summers, F. J., Davis, M., Evrard, A. E. (1995), *Galaxy Tracers and Velocity Bias*, ApJ , 454, 1   93

Sutherland, R. S., Dopita, M. A. (1993), *Cooling functions for low-density astrophysical plasmas*, ApJS , 88, 253   71

Syer, D., White, S. D. M. (1998), *Dark halo mergers and the formation of a universal profile*, MNRAS , 293, 337+   107, 152

Tasitsiomi, A., Olinto, A. V. (2002), *Detectability of neutralino clumps via atmospheric Cherenkov telescopes*, Phys. Rev. D , 66, 83006   149, 161, 177

Taylor, J. E., Silk, J. (2003), *The clumpiness of cold dark matter: implications for the annihilation signal*, MNRAS , 339, 505   123, 149

Theuns, T., Zaroubi, S. (2000), *A wavelet analysis of the spectra of quasi-stellar objects*, MNRAS , 317, 989   33

Tormen, G. (1997), *The rise and fall of satellites in galaxy clusters*, MNRAS , 290, 411   73

Tormen, G. (1998), *The assembly of matter in galaxy clusters*, MNRAS , 297, 648   40

Tormen, G., Bouchet, F. R., White, S. D. M. (1997), *The structure and dynamical evolution of dark matter haloes*, MNRAS , 286, 865   28, 40, 41, 93, 95, 96, 114

Tormen, G., Diaferio, A., Syer, D. (1998), *Survival of substructure within dark matter haloes*, MNRAS , 299, 728   93

Tresse, L., Maddox, S. J. (1998), *The H alpha Luminosity Function and Star Formation Rate at Z approximately 0.2*, ApJ , 495, 691   79

Treyer, M. A., Ellis, R. S., Milliard, B., Donas, J., Bridges, T. J. (1998), *An ultraviolet-selected galaxy redshift survey: new estimates of the local star formation rate*, MNRAS , 300, 303   79

Ullio, P., Bergström, L., Edsjö, J., Lacey, C. (2002), *Cosmological dark matter annihilations into γ rays: A closer look*, Phys. Rev. D , 66, 123502   144, 152, 170

van den Bosch, F. C., Swaters, R. A. (2001), *Dwarf galaxy rotation curves and the core problem of dark matter haloes*, MNRAS , 325, 1017   93

Bibliography

van der Marel, R. P., Alves, D. R., Hardy, E., Suntzeff, N. B. (2002), *New Understanding of Large Magellanic Cloud Structure, Dynamics, and Orbit from Carbon Star Kinematics*, AJ , 124, 2639 129

van Kampen, E. (1995), *Improved numerical modelling of clusters of galaxies*, MNRAS , 273, 295 93

van Straten, W., Bailes, M., Britton, M., Kulkarni, S. R., Anderson, S. B., Manchester, R. N., Sarkissian, J. (2001), *A test of general relativity from the three-dimensional orbital geometry of a binary pulsar*, Nature , 412, 158 19

Viana, P. T. P., Liddle, A. R. (1996), *The cluster abundance in flat and open cosmologies*, MNRAS , 281, 323 24

White, S. D. M. (1978), *Simulations of merging galaxies*, MNRAS , 184, 185 29

White, S. D. M. (1984), *Angular momentum growth in protogalaxies*, ApJ , 286, 38 61

White, S. D. M. (1993), *Formation and Evolution of Galaxies*, in *Les Houches Summer School, 1993* 27, 46

White, S. D. M. (2000), *Building up the Local Group - Theory*, http://online.itp.ucsb.edu/online/galaxy_c00/white/ 94

White, S. D. M., Efstathiou, G., Frenk, C. S. (1993), *The amplitude of mass fluctuations in the universe*, MNRAS , 262, 1023 24

White, S. D. M., Frenk, C. S. (1991), *Galaxy formation through hierarchical clustering*, ApJ , 379, 52 37, 70, 92

White, S. D. M., Rees, M. J. (1978), *Core condensation in heavy halos - A two-stage theory for galaxy formation and clustering*, MNRAS , 183, 341 29, 92

White, S. D. M., Springel, V. (2000), *Where Are the First Stars Now?*, in *The First Stars. Proceedings of the MPA/ESO Workshop held at Garching, Germany, 4-6 August 1999. Achim Weiss, Tom G. Abel, Vanessa Hill (eds.). Springer, p.327*, 327—+ 95

Wyithe, J. S. B., Loeb, A. (2003), *Reionization of Hydrogen and Helium by Early Stars and Quasars*, ApJ , 586, 693 92

Yoshida, N., Sheth, R. K., Diaferio, A. (2001), *Non-Gaussian cosmic microwave background temperature fluctuations from peculiar velocities of clusters*, MNRAS , 328, 669 42

Yoshida, N., Sokasian, A., Hernquist, L., Springel, V. (2003), *Early Structure Formation and Reionization in a Warm Dark Matter Cosmology*, ArXiv Astrophysics e-prints, 3622–+ 21, 94

Yoshida, N., Springel, V., White, S. D. M., Tormen, G. (2000), *Weakly Self-interacting Dark Matter and the Structure of Dark Halos*, ApJ , 544, L87 94, 142

Yoshida, N., Stoehr, F., Springel, V., White, S. D. M. (2002), *Gas cooling in simulations of the formation of the galaxy population*, MNRAS , 335, 762 30, 38, 40

Zel'dovich, Y. B. (1970), *Gravitational instability: an approximate theory for large density perturbations.*, A&A , 5, 84 35

Zwicky, F. (1933), *.*, Helvetica Physika Acta 6, 110 18

Die VDM Verlagsservicegesellschaft sucht für wissenschaftliche Verlage abgeschlossene und herausragende

Dissertationen, Habilitationen, Diplomarbeiten, Master Theses, Magisterarbeiten usw.

für die kostenlose Publikation als Fachbuch.

Sie verfügen über eine Arbeit, die hohen inhaltlichen und formalen Ansprüchen genügt, und haben Interesse an einer honorarvergüteten Publikation?

Dann senden Sie bitte erste Informationen über sich und Ihre Arbeit per Email an *info@vdm-vsg.de*.

Sie erhalten kurzfristig unser Feedback!

VDM Verlagsservicegesellschaft mbH
Dudweiler Landstr. 99
D - 66123 Saarbrücken

Telefon +49 681 3720 174
Fax +49 681 3720 1749

www.vdm-vsg.de

Die VDM Verlagsservicegesellschaft mbH vertritt

Printed by Books on Demand GmbH, Norderstedt / Germany